Praise for My Beautiful Genome

"It's not often you can wholeheartedly recommend a book...
but this is it."   ***Daily Mail***

"*My Beautiful Genome* covers some of the most interesting controversies in biology today, including designer babies, brain
imaging and even whether or not we have free will. It's an
enthralling read."   ***New Scientist***

"A pin-sharp, lively memoir-cum-investigation... Frank's discoveries make for some truly tingling moments... Absorbing."
Mail on Sunday

"The huge research effort to understand the complexity of the
genome is throwing up new insights into the nature of humanity, as the Danish science writer Lone Frank shows in *My Beautiful Genome*, her excellent look into the post-genomic world....
Fascinating."   ***Financial Times***

"Highly engaging... [Frank] writes with wit and good humour
and her analysis is incisive, shedding much needed light on the
tired old nature-versus-nurture debate."
New Internationalist

"Frank is a friendly companion... happily revealing her own
quirks and failings to fuel her search for answers."
Wharf

"A probing biological memoir.... Refreshing [and] wonderfully
poetic."   ***Publishers Weekly***

"I haven't seen Lone Frank's entire genome, but it's obvious from the first page that she's got the SKFF2 gene (Sharp as a Knife and Friggin' Funny, Too). No decoding needed here: I love this book."

Mary Roach, bestselling author of *Stiff* and *Packing for Mars*

"Scientists have deciphered the genetic blueprints of thousands of people [and] narratives about the genomic revolution have not been far behind. Danish science writer Frank's offering is one of the most readable and fascinating of the bunch.... Well played." *Science News*

"*My Beautiful Genome* explores the critical questions and unexpected nuances this new science raises about who exactly we are – as a species, and as individuals."

Brian Christian, author of *The Most Human Human*

"This book manages to do something textbooks miss – to bring home the connection between our genes and our identities... A must-read." *BioNews*

Winner of the Danish Author's Association Prize for Best Nonfiction Book of the Year.

My Beautiful Genome

Exposing Our Genetic Future, One Quirk at a Time

Lone Frank

ONEWORLD

A Oneworld Book

First published in English by Oneworld Publications 2011
This paperback edition published in 2012

Previously published in Danish as *Mit Smukke Genome* by Gyldendal 2010

Copyright © Lone Frank 2010

English translation copyright © Russell Dees

Illustrations by Jørgen Strunge

ISBN 978-1-85168-941-9
eBook ISBN 978-1-85168-864-7

Typeset by Jayvee, Trivandrum, India
Cover design by Jamie Keenan
Printed and bound by Nørhaven, Denmark

Oneworld Publications
185 Banbury Road
Oxford
OX2 7AR
England

Learn more about Oneworld. Join our mailing list
to find out about the latest titles and special offers at:
www.oneworld-publications.com

For my parents — naturally
Irene Frank and Poul Erhardt Pedersen

In Memoriam

The only way to be general is to be deeply personal

Asger Jorn

Contents

My Beautiful Genome

My accidental biology

I'M DEAD TIRED. For the last hour and a half, I've been run through a battery of tests, all designed to shed some light on my personality, my disposition, and my intellectual abilities. I've volunteered to take part in a major research project to examine the connection between specific genes and personality – in particular, a tendency toward depression. We have finally reached the last questionnaire. A young, female researcher is gazing cheerily at me from across a table.

"I'd like to ask you some questions about your immediate family – having to do with drug and alcohol abuse, criminality, and psychological illness."

Her perky blonde ponytail sways back and forth. It makes her look especially efficient.

"They're not about you but about your first-degree relatives: your parents, siblings, and children."

"I don't have any children."

"Your parents and siblings, then."

"My parents are dead, but I have a brother."

"Whether they're alive or not doesn't matter, the questions are the same," she says. "Let's start with alcohol. Have any of your first-degree relatives had any problem with alcohol?"

"Problem? Problem, you say? Yeah, well, I suppose I'd have to say yes to that. Such as it is."

"Yes…?"

"My father. Some would say he had a certain problem with alcohol."

Starting your day with vodka in your coffee and working your way through with malt liquor might be called by some people a bit of a problem.

"For an extended period?"

"As long as I can remember, really. But he didn't think it was a problem himself as such; he could certainly function."

She flips the first page of the questionnaire, following the instructions.

"Did this alcohol abuse ever lead to divorce or separation?"

"Yes."

She looks at me inquisitively, inviting additional information.

"Three times. Divorce."

The eyebrows shoot up her forehead.

"Well, then. Was he ever sent home from work or incapacitated?"

"No, no." Of course not. My father was a very capable and conscientious teacher all his life. He did his job, no matter what.

"No problem there," I reply, thinking the worst is over.

But then she asks, "Were there any arrests or driving under the influence convictions?"

I pause. "A few. That is, I don't quite remember." I feel like I need to explain this, provide a defense. It all suddenly sounds worse than I remember it.

"Nothing ever happened. No accidents, I mean. My father was

an excellent driver, even when he'd had a few. He was just unlucky enough to get caught. A couple of times."

"Okay. Good. So, we're done with alcohol." She resumes the interrogation with a more optimistic tone. "Have any of your first-degree relatives had any mental health problems?"

"Yes," I say without hesitation. I'm asked to identify which ones.

"All of them."

She mumbles to herself, leafing through her papers, confused. "*All of them?* Okay, okay. Where do we start?" I want to be helpful, so I quickly run down the list: When I was little, my mother suffered from depression – deep, clinical depression, which was particularly bad in her last few years. My younger brother has had a few bouts of his own, and my father was manic-depressive, diagnosed at sixty, by which time the disease had come to be known as bipolar disorder.

"He had manic phases?"

"I'd have to say yes." I flash back to that one Christmas when he essentially did not sleep for a week but trudged around the house clutching a stone-age axe in one hand and his well-worn Bible in the other. Talking and talking and talking, becoming more and more incoherent. Finally, we had to hospitalize him.

"Any psychoses?"

Here, I dig in my heels. After all, we're not a family of lunatics.

"*No.* Nothing like that," I reply. "Except, maybe… There were some episodes where my father believed someone was prowling around the garden shed at night to steal his tools. There was also a period when he thought someone was talking to him through the heating pipes, but that was only for a short time. It went away with a little Zyprexa."

She looks down at her notebook again and adds a note. It says "mild paranoia."

"Has anyone other than your father had psychiatric treatment?"

"We all have."

"Medication or consultations with a psychiatrist?"

"Both," I say. Then, something comes to me. "What about suicide attempts, do they count?"

The young researcher nods silently and locates the box on the questionnaire for suicide attempts.

"There were two of them – two that I know about, anyway. Both were made by my father. My mother, on the other hand, talked about it, but never tried it."

The researcher stares resolutely at her papers as she turns to the final questions, having to do with narcotics abuse. Here, I can answer with a clear conscience that no one in my family has ever had any problems with drugs. Never.

"You've never yourself taken narcotics of any kind?"

"I drank some homemade hemp schnapps on New Year's Eve at the beginning of the nineties, but that's all. And it didn't work." Or, rather, it worked so well that I slept through the whole party, which reportedly took place in the great hall of Copenhagen's squatter town Christiania.

"About alcohol," she continues, "I also have to ask you about yourself. How many drinks do you have during the course of a week?"

"It must be around fourteen," I lie, promptly and deftly. For some reason, twenty, or a bit more, doesn't sound good, and my *intention* is always to stick to fourteen. "You know – two glasses of red wine a day, purely for medicinal purposes. It's because red wine contains resveratrol, which is healthy for pretty much anything. Heart, blood pressure, cognitive faculties."

She nods enthusiastically.

"Fourteen drinks, that's within the National Board of Health recommendations. Good, good," she says at last, displaying an almost liberated smile. "Yes, well, I don't think I have any more questions."

BUT I DO. I have questions. They've been smoldering quietly in my mind as we progressed from question to question. They were probably the real reason I volunteered to be a part of this genetic study.

If I am to be honest, there is a direct connection between my interrogation today in a nondescript scientist's office and the hospital room at the other end of the country where I held my father's hand as he died on a summer's day a year earlier. Because what is an interest in genetic information about? It's about your heritage, your history, your identity.

I sat there in that stifling hospital atmosphere with the person I loved more than anyone else in the world, unable to do anything except wait for his end. And when it finally happened, when my father was simply gone in a moment, a single sentence swirled in the back of my head: *I'm an orphan*.

The realization left an icy sensation, not just of being alone, but of being without a source, without a history. Now, there was no one who had been witness to my life back to a time before I could even remember it myself. No one who could see and describe the thread that ran between how I was as a tot and what I later became, who I am today. The past, in its way, was gone. And the future – well, you could see an end to it. At forty-three, I'd reached the age when the chance of having children was pretty much theoretical. That's fine with me, because I'd never seriously contemplated having any, but being both without a source and without any offspring is to be floating free in the vastness of humanity, of life. When you can't see yourself in any other being, you can lose sight of yourself.

Where do I come from? Who am I? Am I going to be like my parents? How will I die? And when?

These are questions humans have always asked, but now they can be asked very pointedly and put to a wonderfully tangible informant – our own DNA. And I cannot help but ask these questions of my biology: I'm a biologist by training. I'm deeply fascinated by the human being as an organism. As the miraculous result of myriad microscopic processes unfolding.

It reminds me of something my father said to me countless times over the years, when he was in a sentimental mood or I needed cheering up for one reason or another.

"My *dear* daughter." There was always a particular emphasis on dear. "You possess an incredibly fortunate combination of genes. You got all the good stuff from your mother and me, but you avoided all the bad stuff." Here, he would embrace a slight pause. "Well, apart from the depressions. But, otherwise, you've got nothing but trophies on the shelves."

What, as a child, do you say to that sort of thing? You roll your eyes and shrug it off. Parental pride is, of course, good for your fragile ego and limping self-esteem, but you also know that it's way off the mark.

"Stop it, Dad, you're talking nonsense."

When I was young, I definitely did not see myself as a slender green shoot topping the stout branches and meandering roots of the majestic tree of my ancestry. I was my own person with my own will, quite independent of previous generations and their idiosyncrasies. What could something as abstract as "biological legacy" mean to me, an individual who was not only perfectly capable of thinking for herself but had no thought but of moving forward? Absolutely nothing.

Now, with my father's death, it's different. Now, it means something. Now, I want to trace my heritage to the roots. To know exactly which genetic variants and mutations have come down to me, and what they mean for who I am. I want to understand how these accidents of biology have shaped my life, my opportunities, and my limitations.

Of course, in front of the mirror, I can see my heritage chiseled directly, and not always entirely happily, in my physical features. The pronounced nose is clearly from my mother's family, where you can spot it back in the sepia-toned portraits of my great-grandfather. My thin, bony frame comes from his wife – my grandfather's crazy mother of whom everyone was afraid. A stingy shrew of a woman with a gift for domestic tyranny, I vaguely remember her from childhood visits to their apartment, infused with the acridity of mothballs and stuffed with heavy mahogany furniture and fussy crocheted doilies. My somewhat elongated, slightly plump face and my narrow lips are clearly a package deal from my paternal grandmother's side of the family tree.

But my familial heritage is not confined to my features. It is undoubtedly from my paternal grandmother's line that I also got my chronic tendency toward sarcasm. Occasionally, I can hear my father's voice in the zingers spurting from my mouth, and I can almost feel his accompanying facial expressions in my features. Is it simply the product of childhood's rigorous social training or is some biology mixed in there? Do we carry this sort of inheritance in our chromosomes? How do nature and nurture collide to create all the stuff that makes people interesting?

"It's not because I like saying this, Lone," said a well-meaning friend at university many years ago, "but your personality is against you." That was around the same time an American friend called me "brutally honest." That judgment made me feel happy about myself until she put her hands on her hips and shouted: "It's cruel! Don't you understand that people despise honesty?"

But how much of the unappealing aspects of my personality can I blame on the minute variations written into my DNA? Do my recurring depressions and consistently dark outlook on life derive from a few unfortunate genes, handed down from two different families? Or do they derive from an upbringing that could at times be, to say the least, challenging?

There is also the issue of physical ailments. I'm not plagued by illness or anything, apart from a touch of rheumatism in the innermost joint of my right big toe, which makes shoe shopping difficult and high heels impossible. But what might be waiting in my future? Will I die the way my parents did? Will I be hit by breast cancer at a young age or be forced to take year after year of pills to regulate my heart and blood pressure? If I took a sneak peek at my genome, could it tell me what is in store for me? And if I know my prognosis well in advance, can I rewrite my future?

—∞∞∞—

WE CAN FINALLY begin asking these questions, because a revolution is under way. Genetics is no longer a matter reserved for scientists and experts; it is becoming quite ordinary, practical, and everyday. In fact, over the next decade, genetics will become as familiar to us as the personal computer. Originally, computers were large, complicated, machines – mainframes – found exclusively in universities and research institutes and only available to the initiated specialist. But then the technological dikes burst, prices fell dramatically, and today computers are the tool of the masses.

But what's the genetics equivalent of the PC? Well, the first genetic dating services are already in business. At GenePartner, based in Switzerland, they claim to be able to match love-starved singles on the basis of selected genes relating to their immune systems. A handful of studies indicate that such genetic compatibility results in both a better sex life and healthier babies. You can also have your prospective boyfriend – this only works for men – tested for whether he has an unfortunate genetic disposition for infidelity or for getting mixed up in bad relationships. If you have children, you can have them tested for whether they possess the genetic disposition for muscles more suited to speed-related or endurance sports. In the next ten years, all newborns will routinely have their genome mapped and deciphered, according to people in the know. And these same technological experts predict that, within a few years, a complete sequence mapping all six billion bases will cost less than the baby's pram.

How can such straight-from-the-womb genome sequencing be used? And will there be any limitations to their real-life application? Jay Flatley, who heads the major league genomics company, Illumina, has argued that "the limitations are sociological," and, of course, he is correct. Social norms and political legislation will dictate what we may do, and culture will dictate our demands and what we actually do.

In China, ambitious, well-to-do parents are already beginning to test the genes of their children before school age, to give them the best upbringing – though whether that upbringing is optimal for the child or the parents is a bit hazy. At Chongqing Children's Palace, a summer camp, one part of the package is a test of eleven different genes that is supposed to provide an excellent picture of your child's potential. The camp's directors send a saliva sample to the Shanghai Biochip Corporation, which returns a detailed statement about the child's intelligence, emotional control, memory, and athletic abilities. This is supplemented with advice from the camp about possible career paths. Is young Jian a powerful CEO in the making, a budding academic, or just a future bureaucrat?

You don't need to send your child to western China if you are anxious to discover and nurture the native gifts of your spawn. You can simply contact the US startup company, My Gene Profile. In their promotional videos, a mustachioed and slightly chubby man explains that good parenting is all about directing your children toward success and happiness, and that this is best done by identifying their abilities through My Gene Profile's test of forty genes. Their test – that is, the interpretation you receive from the company – will reveal the after-school activities for which you should sign up little Emma, and the education that will offer her the biggest pay-off.

Unfortunately, in the here-and-now, this vision of a genetic horoscope is a pipe dream. Both the Chinese children's camp and the American test kit, with its accompanying books on childrearing – available with additional payment – are pure fabrication. Any serious geneticist would shake her head and call them a con or quackery. No one knows of any set of individual genes that can be used to outline a human being's potential or describe the optimal trajectory of his or her life. For now, that is. But the fact that companies can sell this sort of thing says something about the status and role genes have in our twenty-first-century concept of ourselves. It also illustrates the hunger out there to be able to *predict* a life, to shape and optimize it according to our own designs.

Will this ever become a reality? Can the genome be a crystal ball that tells us how life will be? Might DNA be the path to self-knowledge and even a road to change?

I want to go in search of some answers to these questions, and to try to find the limit to which we're willing to probe our futures – *my* future. I want to know how it feels to have a close encounter with my DNA, this invisible, digital self that lies curled up like a fetus in every single cell of my body.

1

Casual about our codons

Get to know your DNA. All it takes is a little bit of spit.

23 ANDME WEBSITE

"THERE HE IS – that's him!"

The man next to me rolls his eyes and tips his head in the direction of an elderly gentleman with an odd bent who is slowly making his way across the lawn beyond us. It's James Watson, the man I've come to this conference at Cold Spring Harbor Laboratory, near New York City, to meet. He's wearing a grass-green pullover and a fire-engine-red bush hat.

"Big Jim!" my interlocutor notes with a broad smile. "If you want to talk to him, you'll have to be aggressive. He's pretty talkative up to a point, but he's become a bit skittish with reporters."

That's understandable. Watson, who together with his colleague Francis Crick, is credited with uncovering the chemical structure of the DNA molecule in 1953, had recently experienced an *annus horribilis*. In 2007, he had run foul of the media machine, and lost some of his luster as a Nobel Prize winner, during a book tour in Britain to promote his latest autobiography, *Avoid Boring People*. In an interview with the *Sunday Times*, he had remarked that he thought the prospects for the African continent were gloomy, because the intelligence of blacks was lower than that of the rest of the world's population. He went on to say he hoped everyone was equal, but that "people who have to deal with black employees find this not true." He also managed to assert that it would be perfectly okay if, on the basis of prenatal testing, a mother-to-be decided to abort a fetus that might have a tendency toward homosexuality. Why not? That sort of choice is entirely up to the parents.

These were opinions Watson had aired in various iterations many times before, but printed in black and white, in a major newspaper, they could no longer be tolerated. Enough was enough. Though a small group of academics defended Watson and tried to explain his statements, the remainder of his book tour was canceled. The Nobel laureate went home to his laboratory at Cold Spring Harbor, where he had been pleasantly and securely ensconced in the director's chair since 1968.

But the furor wouldn't die. Soon after his return, a contrite Watson issued an apology with a built-in disclaimer – it wasn't really what he meant, and blacks are certainly fine people – but the protests continued. Finally, the laboratory's board stepped in. At seventy-nine, Watson accepted an emeritus position and enforced retirement. Not that he was thrown out: he still retains a wood-paneled chancellor's office, in whose antechamber a secretary guards the doctor's calendar with the ferocity of a dragon. And he still shuffles about the lawns and supplements his role as the godfather of genetics with a passion for tennis.

"The most unpleasant human being I've ever met," the well-known evolutionary biologist E. O. Wilson is on record as having said of Watson. And when the old man wasn't being called a racist, the word "sexist" took its place. Watson had a reputation for not accepting female graduate students in his group and for making statements to the effect that it would be an excellent innovation if we could engage in a little gene manipulation to ensure future generations of women are all "pretty."

With this nagging in the back of my mind, I pull myself together and trail after Dr. Watson, notepad in hand.

"What do you want?" he asks nervously. "An interview?"

Watson fixes on me from behind a pair of glasses that make his eyes look like a pair of golf balls. Their assessment does not, apparently, fall in my favor.

"I don't have time," he says softly, turning away. "I have to go home for lunch. I've got guests I need to talk to. Important guests."

He looks around impatiently, as if for someone to come and save him.

"Just ten minutes," I plead, but this time I receive a heavy sigh and a sort of wheeze as a reply. When he remains standing there, oddly indecisive, a kind of impertinence takes over, and I mention almost at random one of the lectures from the day before, on genes and schizophrenia. He's hooked. Watson abruptly draws me inside, into the empty Grace Auditorium, where the conference on personal genomes is taking place, and sits down in one of the front rows.

"My son has schizophrenia," he says. I nod in sympathy — I had heard the tragic story of his youngest son, Rufus. Immediately, Watson begins to mumble. He whispers and snuffles, but his eyes are clear, without a hint of the confusions that often come with old age.

"With respect to genetics, it's still a huge motivation for me to see that this disease is understood. If you ask me what I'd like to see come out of the genetics revolution, it is this: I want to see psychiatric illnesses understood and explained. We have no idea what's going on. Imagine: there are a thousand proteins involved in every single synapse through which a nerve cell transfers impulses to another. And there are billions of them."

As Watson warms to his subject, I run with his change in mood. My own greatest interest, I hurry to tell him, is *behavioral genetics*, in understanding how genetic factors take part in shaping our psyche and personality, our mental capacities, and our behavior as a whole. It is known that heredity is involved not only in our temperament and mood, but in complex matters such as religiosity and political attitudes.

Yet how could a slight variation in the proteins that sail around our brain cells possibly lead to a preference for right-wing or left-wing politics? At one end, you have some strings of genetic information; at the other, a thinking, acting person; and in between, a black box. A box that researchers are only now beginning to prise open.

His golf-ball eyes harden their lock on me. "Mental capacities?" Watson then says in a thin but sharp voice. "Yes, they are interesting, of course – *academically* interesting – but you have to understand that disease is always the winner, when research money is being distributed. And it has to be that way... there are people out there suffering!"

He wheezes again. Whether it is to clear his throat or his thoughts, I am not sure.

"Truth be told, I don't believe there is a chance that the mysteries of schizophrenia will be solved for another ten years. At least ten years."

Many would agree with Watson's assessment. In 2009, a small army of researchers announced the results of three gigantic studies, involving fifty thousand patients scattered across many countries, into the genetic cause of schizophrenia, and they had found very little. In the *New York Times*, Nicholas Wade bluntly called the disappointment "... a historic defeat, a Pearl Harbor of schizophrenia research." The only firm result: no particular genes could be found that determine whether a person develops schizophrenia or not. Furthermore, it is presumably not even the same genes that are involved in all patients.

Here at the Cold Spring Harbor conference, the participants have been discussing the great mystery of genetics: the issue of the *missing heritability*. This is the "dark matter" of the genome. Again, take schizophrenia. Scientists know from countless twin and family studies conducted over decades that the disease is up to eighty percent heritable, but, despite the army of researchers' thorough studies, with tens of thousands of patients, only a small handful of genetic factors have come to light. Altogether, these factors explain just a measly few percent. So where are the rest to be found?

"Rare variants," whispers Watson, as if he were making a confession. "I think it lies in rare variants, genetic changes that are not inherited from the parents but arise spontaneously as mutations in the sick person. Listen: you have two healthy parents and then a child comes along who is deeply disturbed. So far as I can see, it cannot be a question of the child having received an unfortunate combination of otherwise fine genes. Something *new* has to happen. We have to get moving to find this new thing, and my guess is that we need to sequence the full genome of, perhaps ten thousand people, before we have a better understanding of the genetics in major psychiatric diseases."

I ask how it feels to have your genome laid out for everyone to see on the Internet, but Watson pays me no heed. He is lost in his own train of thought.

"Just think about Bill Gates. This man has two completely normal parents but is himself quite strange, right?"

Fortunately, Watson continues before I'm forced to offer a reply.

"No debate," he says. "Bill is weird. Maybe not outright autistic but, at least, *strange*. But my point is that we cannot know in advance who we as a society need. Who can contribute something. Today, it appears that these types of semi-autistic people who are good with computers are really useful. I don't have a handle on all the facts, but I could imagine that in a hundred years, as a result of massive environmental changes and that sort of thing, we human beings will have a much higher mutation rate than we have had up until now. And with more genetic mutations, there will be a greater variation among people and, thus, the possibility for more exceptional individuals."

He gives me a quick sidelong glance. "There are very few really exceptional individuals, and most people by far are complete idiots."

There is a little pause.

"But success in life goes together with good genes, and the losers, well, they have bad genes." Watson stops himself. "No, I'm in enough trouble already. I'd better not say more."

His self-imposed silence lasts five long seconds.

"I mean, it would be good if we could get a greater acceptance of the fact that society has to deal with losers in a sympathetic way. But that's where things have gone wrong – that we would rather not admit that some people are just dumb. That there are actually an incredible number of stupid people."

I suddenly recall one of Watson's classic remarks, that the proportion of idiots among Nobel Prize winners is equal to that among ordinary people. Of course, I don't mention it. That would be crude and insolent, and though I tend to be brutally honest, I know not to push my luck with the old man. Instead, I ask how it feels to be able to look back on the almost incomprehensible advancements that he helped to kick-start almost sixty years ago.

"I never thought that I would have my own genome sequenced – the whole thing mapped from one end to the other. Never. When I was involved in the Human Genome Project, where over several years we mapped the human genome as a common resource, that sort of personal genome seemed entirely utopian. And even when young Jonathan Rothberger from 454, the sequencing company, suddenly offered in 2006 to sequence *my* genome, it sounded crazy. But they *did* it."

The dreamy gaze disappears.

"Today, it's about getting every genome on the Internet, because if you want to know something about your genome, you have to have a lot of eyes looking at it. That's where the money should be going, to get more and more genomes published, so researchers can analyze them and squeeze more knowledge out of the information. Do you know what? They should sequence more old people, because for obvious reasons we are more willing than young people to put our genomes on the Net for public viewing."

Once again, I see the chance to probe Watson about his genome. I want to know how it feels to scrutinize and immerse yourself in your own genetic material. To have made already the journey that I'm hoping to take. "Has knowing your own genome affected you?"

"No, I don't think so. To be honest, I don't think much about it."

"What about the gene for ApoE4?" I ask cautiously. From the beginning, Watson said that he did not want to know whether he has this well-known variation of the Apolipoprotein E gene, which multiplies the risk for developing Alzheimer's disease.

"No, because then I would fret about it, whether it was incipient dementia every time I couldn't remember a name or something like that. Hah! As things stand, I only worry about it half the time."

Since his genome is on the Internet, I wonder whether he really doesn't know the truth or his professed ignorance is pure coquetry. I note out loud that there is no reason to remain ignorant, since he is already eighty-three years old – if he doesn't suffer from dementia now, he probably never will.

"You haven't quite understood," he says, slightly hurt. "Dementia can easily strike in your nineties; it happened to my own grandmother. She was born in 1861 and died when I was twenty-six. Wonderful woman, by the way. I must tell you… " He turns around to face me. "I know many men in their eighties who are still razor-sharp, but I don't know many role models who are over ninety. Something happens to most men between eighty and ninety."

For a moment, I think that he has revealed a sense of humor, but before I laugh I catch the expression in his eyes. He is utterly serious.

"But there is something else. I believed that, as a white male of European origin, I could tolerate milk and have always drunk it. And I've eaten ice cream, lots of ice cream. But my genome reveals that I am partially lactose-intolerant. Today, I only drink soy milk and must admit that I actually have fewer gastric problems."

This was, perhaps, too much information.

"Everyone should have this kind of knowledge from birth, so mothers could make sure their children have the best possible diet." He muses on another example: heart attacks and hypertension. "I also have a gene with reduced activity that makes me metabolize beta blockers poorly. Because I have high blood pressure, the doctors had already given me medicine. With this genetic knowledge, it was suddenly not strange that the pills just made me fall asleep. One out of every ten Caucasians has a genetic variation that makes beta blockers ineffective for them. Everyone should be screened for that kind of stuff, right?"

Suddenly, Watson changes track.

"We've reached a point where we have to ask ourselves how much we can defensibly contract out to private companies. I'm an academic, dammit, and I would rather see my friends' genomes mapped by an academic lab like the Broad Institute in Boston or the British Sanger Centre than by some company. These outfits come and go and have no real interest in the science," he says, staring vaguely in the direction of the larger-than-life-size portrait of himself that is the auditorium's only decoration. The artist looks to be an admirer of the British painter Lucian Freud and, like his role model, has depicted his subject with every fold of skin and liver spot.

The living Watson slumps a bit in his chair. He looks infinitely tired, an old turtle more than an old man. He shakes his head.

"I don't know where we end. Think that we've come so far that everyone can not only have his genome sequenced but can have it done by Google."

He relaxes his arms over the chair and falls into a reverie.

"Are you going back to Denmark now?" he asks out of the blue, for the first time in a directly friendly tone. I answer in the affirmative.

"Poor girl. That country is the saddest place I've ever been. Before I went to the University of Cambridge, I was there for a whole year, doing virus research for the National Serum Institute in Copenhagen. As I remember it, the sun never came out once."

---⊗⊗⊗---

THE YEAR JAMES Watson spent in Denmark was three years before the breakthrough that transformed him forever into Big Jim and revolutionized humanity's understanding of biology. As his colleague Francis Crick shouted that evening after the two had finally grasped the helix form of the DNA molecule and tumbled into the Eagle Pub in Cambridge: "... we had found the secret of life."

In his classic account, *The Double Helix*, Watson describes how he hit on the structure one day when he was sitting in the lab playing around with cardboard models of the four bases that serve as the genome's universal building blocks: Adenine, Guanine, Cytosine, and Thymine. At once, the young but intensely ambitious researcher realized how these bases must fit together in fixed pairs: that A and T combine with two weak hydrogen bonds, while G and C bond with three. It also became obvious how the bases had to face each other and, thus, hold the molecule's backbone – the two long phosphate strands – in a three-dimensional double helix. A beautiful, biological, winding staircase.

Until that moment, Watson and Crick had been engaged in a long and merciless race. On their perch at the University of Cambridge's Cavendish Laboratory, they had been out in front, but the legendary Nobel Prize winner Linus Pauling, of the California Institute of Technology, had nearly caught up. It was difficult for most observers to believe that Pauling would not win the contest, but the old giant became mired in some blind alleys involving his work with proteins. That allowed Watson and Crick to cofound the genetic revolution.

To have hit upon the double helix structure was equivalent to breaking through a wall – a thick, reinforced concrete wall. Now they had discovered not only that the previously mysterious DNA was the bearer of an organism's genetic heritage, but also how the whole thing was screwed together chemically. The molecule's structure was crucial. Only once it was revealed did scientists finally have a key for unlocking the genetic mechanism's operation at the most basic level: the manner in which characteristics that develop over the lifetime of an individual can be physically passed on from generation to generation, nicely packed into an egg and a sperm.

On the surface, that process almost seems like magic, like pure mysticism. In the nucleus of all cells are some genes – heritable units, so to speak – that are themselves unchanging and static but which nevertheless provide the source for the eternal change and dynamism that characterizes every living organism. A human being's genome – his or her total hereditary material – consists of forty-six different chromosomes, each of which is one long DNA molecule. These comprise the two sex chromosomes, X and Y, and twenty-two ordinary, housekeeping chromosomes that are each supplied in two different copies – one from each of our parents. In a way, the genome behaves like a queen ant. She remains passive, hidden, and protected deep in the center of the colony, where she is serviced by diligent worker ants and from which, via her production of different types of offspring, she controls the life of the entire ant society. Correspondingly, the genome is found in the cell nucleus, from where its information is read and transmitted to the rest of the cell, and to the organism, through a series of molecular middlemen.

The miracle is that along the way an elegant and gradual transformation from a digital code to an analog reality takes place. Genes do not do anything; they just are. But the information they contain – the genes' essence, as it were – is converted into the stuff that realizes biological ideas – namely, proteins. Proteins are the workhorses of the organism. These large, clumpy molecules with their moveable parts and biochemical capacities can carry out all the tasks life requires.

We are not only largely built of protein, which is found in every cell and organ structure, we also function via proteins. Enzymes are a specialized class of proteins that take care of our biochemistry, and another class of specialists are the receptors – proteins that are responsible for all sorts of internal communication, through conveying chemical signals within cells and between cells and organs. In short, there are proteins in everything, and every single one of them is built on the basis of information in a corresponding gene.

The process of changing from gene to protein is a precisely choreographed dance. Each of our forty-six chromosomes is a long, unbroken DNA molecule. Imagine a long spiral, like a zipper, the teeth of which on each side are the simple bases A, G, C, and T. When a protein is produced, the zipper is opened in the place at which the gene is found, and special enzymes start transcribing a copy of the information. The copy is produced in RNA, which is a structural cousin of DNA with slightly different chemical components.

The small molecular transcript from a gene is called a *messenger RNA* (mRNA), and it is just that – a messenger. It is dispatched from the compact nucleus out into the cell, where it offers itself for

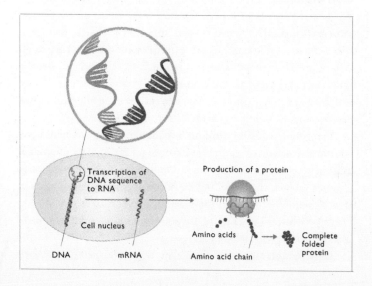

Transcription of DNA sequence to RNA

Production of a protein

Cell nucleus

Amino acids

Amino acid chain

Complete folded protein

DNA

mRNA

translation. The translation occurs in large protein factories, which are themselves built from a series of proteins and which read messenger RNA molecules as recipes.

The recipe for a protein is itself in the genetic code. A given sequence, that is, a sequence of bases in the RNA molecule, specifies one and only one corresponding sequence of amino acids which, when they are linked together, constitute a protein. A genetic sequence of three bases – a codon – specifies one and only one amino acid out of the twenty that organisms make use of. If you have a genetic sequence CCC followed by AGC and then ACA, it means that the amino acid proline is to be hooked up with serine, which in turn is to be connected to threonine.

Correspondingly, there are start and stop codes. A messenger RNA is translated into protein by the code being continuously read and translated into a growing chain of amino acids. When a stop code appears, the process completes, whereupon the finished protein is spit out into the larger, complicated cellular machinery with specialized compartments where further modifications and renovations are made.

Nearly every one of an organism's cells contains basically the same information hidden in the genome. They each have their own special identity, due to the fact that the information inside is treated differently. Only those genes that are suited to the individual cell's tasks are read, translated, and allowed to produce protein. A liver cell forms protein from a certain array of genes, while a brain cell uses a completely different array.

In the midst of all this reading and translating, genetic mutations occasionally occur. Mutation means change, and genetic mutations assume many forms. There are point mutations in which one base is changed into another base, and there are larger changes that either remove a number of bases (*deletion*) or add new ones (*expansion*). Finally, pieces of the DNA of chromosomes can be rotated so its base sequence is inverted.

Genetic mutations can change an organism's proteins in sev-

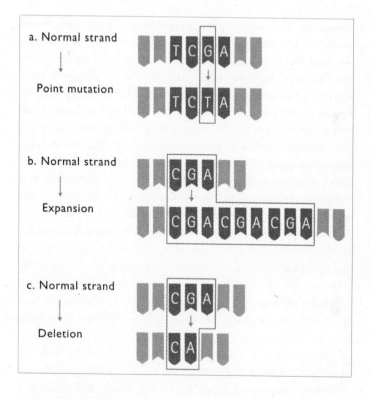

a. Normal strand
↓
Point mutation

b. Normal strand
↓
Expansion

c. Normal strand
↓
Deletion

eral ways. A single point mutation can replace an amino acid in a protein, thereby making the protein fold in a new way and, perhaps, making it more or less effective for the tasks it performs. Larger mutations can likewise change the function or completely inactivate the protein. Finally, mutations in DNA regions that do not themselves produce protein but, on the contrary, regulate production, can give rise to the formation of more or less protein.

These changes can trigger physiological effects, some of which are beneficial and some of which are harmful. Mutations arise all the time because of damage to the genome or mistakes in DNA copying, and these mutations, which are passed on to the next generation, may over time be dispersed in the broader population or disappear again. They are, so to speak, the fuel of evolution.

We all have the same genes, each in two copies, one from each parent (apart from the single genes of the sex chromosomes). However, because of the mutations that have occurred over the course of evolution, millions of which have survived and spread throughout the species, most genes exist in different variants. The astronomical number of possible combinations of these variants means every one of us is physically and physiologically different. Even identical twins. Although the twins' original genome sequences are identical, mutations, and especially individual epigenetic modifications, accumulate throughout life, differentiating their genomes. Each of us carries a unique, beautiful genome.

NOT UNTIL 1963, ten years after the DNA structure was served up by Watson and Crick, was the genetic code – the language of genes – finally cracked. And with very few exceptions, that language is universal across Earth-bound life. Whether you are a flu virus, a slime mold, a manatee, or a manager, your genetic code contains the same components. From this scientists derived another piece of knowledge:

> *Life is not based on chemical substances or molecules but on information, pure and simple.*

Well, "*duh*," we say today with a shrug. And without raising an eyebrow, we can recite the statistics that human beings share ninety-eight percent of their genome with a screeching chimpanzee, sixty percent with a skittering mouse, and even twenty percent with a lowly roundworm a millimeter long. But take a moment to think about this, slowly and carefully. This insight goes deep, and touches on something central, something almost psychologically jarring.

For one thing, it testifies to a common and global biological heritage that is not superficial, but reaches into the very core of all living creatures.

For another, it forces us to think about life in a new way. The phenomenon of life should not be viewed as a number of fixed, defined forms – slime mold, manatees, managers. Rather, it is a continuous stream of information. The myriad specific life forms are just temporary vessels holding the genetic information before it is transmitted on and on through time in novel combinations.

It also allows us to consider biology in terms of the digital world. Genetic information is like software programs and data that are expressed in binary code and can be read in the same way by different computers, whether a monstrosity of a stationary IBM, a sleek little Mac or, for that matter, a mobile phone. The genetic code's message is the same. In a corresponding way, a brain cell in a human being reads and translates "gene language" in the same way as a yeast cell.

The significance of this realization stretches far beyond the psychological. Biology's fundamentally digital nature has a dramatic consequence: genetic information does not belong to a particular place but can be freely transplanted between very different organisms. There is nothing particularly "rose-like" about a gene found in a rose; the same gene can just as easily produce its protein in any other living organism.

This knowledge became tangible in 1973, when the molecular biologists Stanley Cohen, Herbert Boyer, and Paul Berg showed they could move information between organisms. By using naturally-occurring enzymes that clip and cut DNA, they showed that a gene cut from the skin cell of a frog could be transferred to a bacterial cell and that, undaunted, the bacterium proceeded to produce the protein specified in the frog gene. In other words: a gene is a gene is a gene.

Thus *gene splicing,* also known as gene technology, was born, and science kicked down the door to a new world, in which genetic information could be moved freely between individuals and even between species that could never have mixed their genes in nature. New forms of life were imagined, from plants with practical traits designed specifically for agriculture, to microbes engineered to produce medicinal proteins. Incredibly clever, all of it.

The scientific community's high enthusiasm was matched by its high anxiety. What unimagined life form might pop up along the way? Could the manipulation of nature's creations wind up disturbing or destroying the complicated ecological puzzle that billions of years of evolution had created, honed, and fine-tuned? Were we playing with Pandora's box?

To tackle such questions, in 1975, the top researchers of the time gathered together in California, at the now-legendary Asilomar Conference on Recombinant DNA. They did not just want to discuss how gene technology could be pursued with the greatest possible safety; they also wanted to jumpstart a public debate on the subject. It was just after the Watergate scandal, and there was a general longing for transparency in societal institutions. Now it was science's turn to come out of the closet. One faction believed that the only sensible approach was to impose, for a defined time, an immediate moratorium during which everyone desisted from gene technology experiments while they considered what the consequences might be. Another faction argued that guidelines should be immediately established, and experiments should be run within defensible safety margins.

The latter group won the day. And with their victory, molecular biology reshaped the entire field of biology. Within a couple of decades, work on gene technology dominated biological research. Today, there is not a single sub-discipline in biology that does not involve genetic knowledge or genetic data. Even a botanist must occasionally take off his rubber boots and log onto a database. Kinships between plants are no longer determined by fiddling with petals or reproductive organs but by comparing genetic sequences among species.

The list of organisms that have been subject to gene manipulation is long, almost boundless. Strawberries have been pumped full of antifreeze proteins, which they produce via a gene borrowed from a deep-sea fish. Genes that code for a spider's silk proteins are placed in yeast cells, which then happily produce great quantities of fine spider silk that can be spun into super-strong, flexible materials. Christmas trees are made resistant to disease, and flowers are equipped with colors they would never have discovered themselves. Innocent aquarium fish have been given a gene from a jellyfish that makes their bodies glow fluorescent green. And pigs are given genes for human diseases, such as Alzheimer's, so that they can serve as model organisms for research into a cure.

Manipulation is one broad lane in the gene research highway; mapping is the other. We are living our version of the Enlightenment's obsession with taking a census of the planet's manifold life forms, describing them, comparing them, and thus creating order in their mutual kinship. Now, genetics is the key to all order and understanding. One organism after another has had its genome sequenced from one end to the other, and the sequences have been entered onto massive databases.

It began with the smallest life forms, viruses, which are not actually alive, but rather "parasites" built exclusively of genetic material. Then came "real" organisms, from bacteria and molds to plants and animals — almost four thousand organisms, including *Homo sapiens*, from every step in the evolutionary ladder.

"The most wondrous map ever produced by humankind" is the map of the human genome. It began in the 1980s, as a crazy idea among the most far-sighted geneticists, who envisioned a tool that would unravel the inner workings of our biology and speed up disease research considerably. The project was monumental for the technology then available, and the Human Genome Project, an international consortium, was established to get it up and running. James Watson, by then no longer a lanky boy with big ears and sharp elbows, but a power to be reckoned with in the research world, was one of the driving forces. The consortium would piece together the entire human genome, a sort of encyclopedia of us. When the project was officially launched in 1990, it was expected to take fifteen years to compile this great reference work.

For years, mapping proceeded apace without much ado. Scientists at the Sanger Institute, near Cambridge, UK, and the National Institutes of Health near Washington, DC, among other laboratories, steadily worked away. Then, out of the blue, war broke out. The American geneticist and entrepreneur, J. Craig Venter and his for-profit company Celera promised to map a complete genome faster and cheaper than all the academic groups combined. Venter had established a veritable sequencing factory in Rockville, Maryland, just down the road from NIH, and filled it with a new generation of sequencing machines and a phalanx of supercomputers rigged to put together the genetic jigsaw puzzle. The initiative was not welcomed. Venter was called a maverick in the newspaper headlines, and Watson, in his typical style, came out and called his competitor "Hitler."

The acrimony ended with a compromise, in which Celera and the Human Genome Project exchanged vital data. This convenient union worked together to produce the first rough outline of the human genome in 2001, four years earlier than planned. The sequence was launched with grandiose rhetoric. The US President, Bill Clinton, and the UK Prime Minister, Tony Blair, appeared hand in hand on television, speaking of the "Book of Life."

But the book contained some surprises. Researchers were stunned at how little content there seemed to be. The scientific consensus had predicted that there must be around a hundred thousand genes in the human genome, but the analysis indicated that it was closer to twenty thousand to thirty thousand – shockingly few.

Genes constitute only about two percent of the whole genome, appearing like pearls threaded on a long string. Each gene corresponds to a sequence that twines between a start signal and a stop signal; in between the genes lies a sea of DNA that never produces protein. The remaining ninety-eight percent of the genome is a mixed bag of components that are, as yet, only poorly explored and understood.

The genes themselves are flanked by regions that help regulate how active the gene is. You can compare it to a speed regulator, which can boost the production of RNA or gear it down, according to need. But these regulatory regions don't take up much space, and most of the rest, by far, is what is sometimes called *junk DNA*, simply trash. A part of it – in fact, eight percent of the total genome – looks like nothing more than a virus graveyard, containing sequences stemming from a motley crew of viruses that, at various points far back in the depths of evolution, have latched on as stowaways and lost their ability to make us sick.

Other "junk," though, is beginning to reveal its genuine function. It turns out this mess is not just passive – it gives rise to huge quantities of RNA molecules that are never translated into protein, but travel around the cell on their own. Scientists have already discovered that some of these RNA molecules help regulate how active "real" genes are; others are still waiting to have their role clarified.

So, from the genetic material of a few hundred anonymous individuals, hundreds of thousands of lab-hours, and four billion dollars, the Human Genome Project gave us the first biological map of humanity. The technological developments that made it possible also now mean that things are moving at an incredibly rapid pace. Today, the cost of sequencing DNA is almost in free fall, to the point where the fall is faster than would be predicted by Moore's law (which says that the price of calculating power of microchips will drop fifty percent every year and a half). Between 1999 and 2009, the price shrank by the factor of a whopping fourteen thousand, and in 2010 – barely a decade after the first genome was sequenced – companies such as Illumina and Complete Genomics could sequence a person's genome for six thousand dollars. The work is done by a single machine and takes a single day.

More than anything, the genome project is the emerging seedling of a genetic-industrial complex. Once the initial vision was fulfilled, the international machinery of discovery that had been established at large research institutes could not just stand idle, and scientists pinpointed new projects to throw into the works. They shifted their focus from creating a map of our basic, common genetics to investigating individual variation. The goal was to catalog all the differences found in our DNA because they, of course, are central to understanding the differences between people.

The first big enterprise, HapMap, aimed to throw light on the variation between the large geographic groupings we traditionally call "races." To see if these groupings correspond to actual genetic differences, HapMap mapped the genomes of people from five major ethnic groups. The effort yielded the first catalog in which *point mutations* – substitutions of one base for another in the genome – appear. These commonly studied mutations have gained an idiomatic abbreviation, "snip" (SNP or *single nucleotide polymorphism*). It is estimated that there are somewhere in the neighborhood of fifteen million SNPs in the human genome, though just over three million have been identified and conscientiously entered into the international SNP database.

Future steps will involve cataloging variation in ever finer detail, by sequencing as many complete genomes as scientists can get into their labs. At the end of 2010, the list of published, freely accessible genomes stood at two dozen and included the genomes of celebrities such as Bishop Desmond Tutu and actress, Glenn Close. Research institutions hold unpublished records of almost two hundred more people. But that's only the beginning. The 1000 Genomes Project is sequencing the genomes of over a thousand volunteers for public use, and the even more ambitious Personal Genome Project is endeavoring to collect a hundred thousand genomes. These genome collectors are pursuing all types of variation.

We humans do not just carry around small, discrete SNPs, with an A in one and a G in another, ready to be noted and entered on a database. We also carry far more drastic mutations: large pieces of DNA may have fallen out, been duplicated one or more times, or simply have moved around and landed in different places in the genome. For the mutation hunters in the research community, the preferred tracking tool of recent years is the so-called *genome-wide association study*. Association studies are supposed to highlight which genes influence certain diseases or other human characteristics. They are incredibly simple in their design. A group of volunteers suffering from a particular disease, or possessing a particular trait, is compared to a control group that does not have that disease or trait. Both groups are tested for a number of known SNPs, using a "gene chip." Inside this ingenious, postage-stamp sized device, small pieces of DNA are placed, which flutter like sea grass on the ocean floor. Current gene chips can typically test between half a million and a million SNPs, drawn from across the landscape of chromosomes. This massive quantity of data is fed directly into a computer, and software is used to find patterns in the chip's genetic observations.

The question posed by such bioinformaticists is basic: does one SNP (or more) recur far more frequently in the sick subjects than in the well ones? If it does, the mutation is considered a *marker* associated with the disease. Once scientists know where the marker is to be found in the genome, they can use its placement to work out which gene it is associated with. They can also calculate how much the presence of the marker increases the risk of developing the disease. This method makes it possible to study the causes of a disease without first forming a hypothesis of where the problem resides. It's the equivalent of casting a large, finely meshed net across the genome and seeing what comes in with the catch. As soon as the bioinformaticists and their computers have hit on an association, the biologists can run to the lab and investigate the identified gene. By clarifying the gene's role in the organism's tissue, researchers hope to gain insight into the biological mechanisms of the disease and, eventually, discover new treatment methods and drugs.

One early, and very well-known, association study whetted appetites in the field. In 2005, a team led by Robert J. Klein of Rockefeller University was able to show that a serious eye condition, age-related macular degeneration, is clearly associated with a SNP variant in a very particular gene already known to produce a protein that regulates inflammation. The result killed two birds with one stone: not only was there a connection that could be used to identify people with a high risk of getting the disease but also a threshold that could be used to clarify its mechanisms.

Two years later, the first truly significant association studies were published in the journal *Nature*. In the first, scientists from McGill University in Montreal tried to locate the genetic factors involved in type 2 diabetes, one of the major lifestyle-associated diseases of our time. After testing nearly four hundred thousand SNPs, the McGill group found many associations – notably variations in two already well-documented genes. Later that year, the Wellcome Trust, in the UK, backed similar studies for type 1 diabetes, type 2 diabetes, high blood pressure and cardiovascular disease, rheumatoid arthritis, Crohn's disease, and bipolar disorder. Two hundred researchers analyzed data from seventeen thousand people, healthy and sick, and revealed a number of new genetic associations.

To date, association studies have provided more than four hundred associations between particular gene variants and everything from prostate cancer, to kidney stones, to curly hair, and even to something as esoteric as the ability to smell digested asparagus in urine.

NICE. BUT CAN anything practical come of finding the gene that makes some of us super-sensitive to après-dîner asparagus fumes? It happens that the ever-cheaper technology for sequencing genes, studying genetic variation, and running association studies has given birth to a brand new creature living beyond the laboratory. As the well-known American psychologist Steven Pinker strikingly remarked: "We have entered the era of consumer genetics."

The Great Leap Forward to consumer genetics came in 2008, when the masses were finally invited to join James Watson, Craig Venter, and a handful of celebrities, at the sequencing party. In a heated race to be the first to the market, two companies, the deCODEme, in Iceland, and the 23andMe, in the US, began selling personal gene profiles. If you mailed in a bit of saliva or a cheek swab, deCODEme or 23andMe would test your sample, searching for between half a million and a million genetic markers. Your SNPs would then be compared with results from several high-profile association studies – checking to see if your genes match those associated with, for example, cardiovascular disease, diabetes, or Alzheimer's disease. Association studies are not cut and dried, however. You aren't receiving a genetic diagnosis, informing you that you have or are very likely to get a genetically determined disease. On the contrary, you're getting a risk assessment, a collection of indicators that compare your chances of getting a disease to the general population's:

> *According to your genetic profile, dear customer, you have an 8.7% risk of developing disease X at some point in your life. This risk is twenty-five percent higher than the average for people of your ethnic background.*

With your genetic risk assessment in hand, you can devote yourself to goal-directed prevention. Like Chinese parents sending their children off to genome-testing summer camp, you can study your genetic makeup and work out what you should do to optimize your health and longevity, if not your career prospects. To make the process as technologically efficient (and perhaps cost-effective) as possible, everything is served up virtually, without human contact, via the Internet in a format in which you can not only browse your own genome but invite your "friends" as well.

This is genetics as Facebook. Information that has always been hidden and unobtainable is now visualized and displayed, and what was broadly considered ultimately personal is now readily shared and compared. Presently, we're in the Wild West of personal genetic services – young, exciting, and full of golden opportunities. In fact, in just its first few years, the scope of the industry has grown beyond the comprehension of the average consumers it is trying to lure. There are somewhere in the neighborhood of 1500 distinct genetic tests in which mutations in a single gene are mapped, and almost just as many businesses that offer them.

Beyond those who are peddling tests, there are websites where you can orient yourself with your genetic results, and advice bureaux that offer to take you by the hand and, with the help of some questionnaires, identify the genetic tests and package deals best suited to your needs, exactly as if you were considering a cruise holiday, a flat-screen television, or a washing machine. As in any booming industry, the genetic supermarket is stocked with ever more goods each day. And the store is open twenty-four/seven.

As we get used to shopping, our genetic interest in ourselves and each other will reach widely and in many directions. It may well be that James Watson is right, and disease research will have the highest priority and get the most funding. But as the price of genetic testing falls and entrepreneurial opportunities rise, the demand for studying normal traits will become insistent. We will put the spotlight on behavioral genetics, that once-controversial field that asks the cheeky questions about how tiny differences in individual genes can explain how people think, react, and act differently. Questions that go to the very heart of human nature.

That should not be surprising. Our DNA is, in its way, our most intimate history. In the future, we will read the development of our species through comparisons of our genome with those of other species, and trace human kinships far and wide as they are inscribed in the alphabet of the four bases. When we want to, or need to, we will seek out our identity and our affiliations directly in our genes.

2

Blood kin

Being born in a duck yard does not matter, if only you are hatched

from a swan's egg.

HANS CHRISTIAN ANDERSEN

"EXCUSE ME, FRÄULEIN, but are we related?"

I'm not sure what the man at the counter is saying. I'm exhausted after the trip from Cold Spring Harbor, New York, to Frankfurt, Germany, and I've simply asked him for a window seat on the plane to Copenhagen. Why is some random airline clerk asking me about my relatives?

"Frank," he says, pointing to his nametag. "My name is Frank, too. Eberhard Frank." He persists: "I come from Pomerania; could we be related?"

I look at him but can't see any particular resemblance. We're both pale and have the mousy-brown hair typical of many northern Europeans, but then, it's typical because half the population at these latitudes has it. What can I say – I've never been to Pomerania and have no idea whether I have distant German relatives there. I settle on a moderately amicable, "I doubt it."

When I think about it, I realize I know embarrassingly little about where I come from and have not previously given it much consideration. Like most people, I'm familiar with about three generations of my family tree: my parents, my grandparents and my great-grandparents – or, rather, some of my great-grandparents. I was lucky enough to meet two of my great-grandmothers and one great-grandfather at an age when I was capable of forming memories of them. The others died before I was born. Beyond that, everything is lost in a fog. Ask me about my great-great-grandparents, and I could tell you nothing.

Before I started on my quest to probe into my genome, I never had a sense that kinship – *genetic* kinship – was especially important to identity. Obviously, your close family is important, because you have a personal relationship. But whether I shared DNA sequences with people I did not know or a throng of people long dead had seemed to me emotionally meaningless. Almost hypothetical.

"I am what I *do*," I always say, emphatically. My identity is my work. Where is the logic in defining oneself in terms of some randomly transmitted molecules?

Still, you can't get around it. There is something very powerful and identity-producing in our DNA. Biology – bloodlines – *means* something, even if it might not have much to do with your daily routine or your social network. It reminds me of A.M Homes's observations on biology and culture in her memoir *The Mistress's Daughter*. As an adult, Homes, who is adopted, meets her biological parents – a couple of opportunists whom she finds she does not actually like. She becomes suddenly and enormously interested in exploring their roots, and thus her own. Homes writes:

> I note to myself that I am not as interested in digging into the story of the mother and father I grew up with, and I don't really know why. Is there something physically unique about discovering this new biological tale?

To her surprise, her biological family seems more real than the adoptive one in which she had lived since just a few days after her birth. Blood mattered.

That may have something to do with our time. We are served reality television programs which transport celebrities to musty archives or distant lands to unearth a biological ancestor. In other, equally schlocky programs, adopted children learn their "true" past, locating distant relatives or ancestors with whom they share neither a language nor a culture. We are blasted with stories about sperm donors, as children of anonymous donors demand to know more about their biological parents, while some donors – such as the English firefighter Andy Bathie – have been held responsible for child support on the basis of biology alone.

Of course, genetics is about family, but it's also about finances. One of the most distinct trends in consumer genetics is the increasing use of paternity tests. There have always been fathers who wondered whether the redheaded son with the odd nose could really be theirs, but today they can gain absolute certainty about a wife's fidelity by purchasing an easy-to-use DNA test. In Denmark, five thousand paternity tests were conducted by private firms in 2009, a fifty percent increase in a single year. One UK company suggests that something of the order of seventy thousand paternity tests are made annually. In the United States, it is estimated that up to a half a million tests (provided by officially controlled laboratories) are conducted each year just for use in lawsuits: the number of private tests is presumably much higher.

Since 2008, it has become possible for anyone to do a paternity check simply by buying a commercial test off the shelf at the local pharmacy. The test is relatively pain-free – depending on the results. A man suspicious of his wife's moral fiber can take the swabs supplied with the kit, scrape some cells from the inside of his and his child's cheek, and send them off by mail. Sometimes the answer is: "Unfortunately, comparison of the DNA samples submitted cannot confirm paternity." And what then? If he's raised a child as his own, changed the diapers, helped with the 4 a.m. feedings, nursed the chickenpox, and coached through the multiplication tables, can the child really stop being his because they do not share his genome?

Aren't the feelings cultivated through years of living together worth something?

Not necessarily. The *New York Times Magazine*, for example, has described how accessible DNA tests are changing the notion of fatherhood itself, as a conflict has arisen between the traditional view of fatherhood and a strictly genetic view. Under US law, with respect to your obligations as a child's provider, it doesn't matter whether or not you are biologically related, but more and more "fathers" are refusing to pay child support, sometimes even abandoning children, when they discover there is no genetic kinship. The argument seems to be *Why spend time and money on a relationship in which you have no biological stake?* It's not easy for the children, who lose a parent in the battle over bloodlines; they are usually unprepared for the redrawing of kinship ties. In the article, a teenager, called L learned as a nine-year-old that she had a different father from the one she had lived with all her life. "At first, it made me scared, because if my dad wasn't related to me, then I was living with someone who wasn't a part of my family, like a stranger," she explained.

Biology isn't limited to these most intimate of family relationships. In a far more diffuse form, genetics also creeps into what you might call the extended, or hyper-extended, family – that is, one's ethnic group, one's people. The first time I was personally confronted by this was in the 1990s, when I was living on Manhattan's Upper West Side in a venerable old building on the corner of 112th Street and Broadway. There, I was always thought to be Jewish, because my last name *could* be Jewish. Think, for instance, of the immortal diarist, Anne Frank.

The Upper West Side is a heavily Jewish neighborhood, where you are never far from a synagogue. On Friday evenings, I would see men wearing curled side locks and plain black clothes plodding up and down West End Avenue, trailed by wives who covered their hair with pious stiff wigs. People here took their ethnicity seriously. I didn't know a soul in the neighborhood, but a few days after my name was placed on the apartment door, unsolicited pamphlets and leaflets bearing the Star of David and menorahs came pouring in.

There were invitations to events at the temple, and letters encouraging "Dear Ms. Frank" to join various Jewish organizations as quickly as possible. One afternoon, I ran into a diminutive older gentleman who was about to shove a pink flyer through my letter slot. I told him politely that it was a waste of time. So far as I knew, my name did not derive from Jewish ancestors. In fact, the closest I'd ever come to anything Jewish was the six months I spent as a volunteer in an Israeli kibbutz right after school – and that was more for the booze and the parties.

"That's too bad, but of course it's not your fault," intoned the man sympathetically, placing a hand on my arm.

Soon after, the stream of paper dried up, but I again encountered the special connection that genetic ancestry creates – with a slightly different approach. I had met a few Jewish men who were attracted to non-Jewish women but who admitted flat out that they would never dream of marrying or having children with them. None of them could give me a good explanation as to why. When I pressed them, in my direct and brutally honest way, they said it didn't have so much to do with culture – that's something anyone can learn – or with religious faith, which none of them had. The closest I got to an explanation was when one mumbled, "there's just something important about the biological connection to the past." For him, it seemed, genes were the essence of a human being.

You can observe a similar sentiment in recent years among African Americans, who have used genetic tools to trace where their forefathers came from. For them, it is not enough to know the broad historical account of the slaves hauled in their millions to American coasts, where they mixed – and were mixed – with other groups and created a new, common culture. No, they wanted to connect with a specific line of origin – a country, an ethnic group, or, better still, a particular tribe in a particular village. Despite the fact that many of these Americans have never visited the African continent, they identify strongly with a distant biological affiliation with what is today Ghana or Côte d'Ivoire, though neither country existed at the time of the slave trade.

Some of the interest in tracing genetic roots back to Africa has been stimulated by celebrity DNA tests. Chris Rock, the comedian, sought the counsel of the genetic oracle and learned that his ancestors hailed from Cameroon and the Udeme people. Based on a DNA test she took in 2007, Whoopi Goldberg can brag that she is a descendant of the Papel people of Guinea-Bissau. (The country's tourism ministry invited her to visit, but she declined.) And Oprah Winfrey threw herself into building girls' schools in South Africa after she discovered in 2005 that her DNA showed her to be a descendant of the region's Zulus. Suddenly, the fate of the local population lay heavy on her mind. Alas, the connection to the Zulus turned out to be a mistake. Winfrey was tested again, and this time proclaimed to be a descendant of the Kpelle people in what is now Liberia. The extent to which this will benefit Liberian schoolgirls remains to be seen.

Such tests are the primary business of African Ancestry, which promotes itself as the only company with the technology to trace a person's heritage back to a country and a region in today's Africa. The company, which makes a point of being "black-owned," claims that its information edge comes from a database of twenty-five thousand DNA samples taken from the people of thirty African countries and covering two hundred ethnic groups. For $349, African Ancestry compares selected genetic markers in a customer with the markers from Africans in its database and – with luck – finds an ethnic and geographical match.

Academic specialists, including the geneticist Deborah Bolnick of the University of Texas at Austin, have raised serious doubts about whether African Ancestry has enough data to satisfy its advertised promises. But that doesn't slow the stream of new business – presumably because customers are not seeking the absolute truth but instead an identity, and are willing to think of that identity as an ongoing project. As the founder of African Ancestry, Gina Paige, told the BBC, the real goal is to change the way African Americans look at themselves.

BUT WHERE DOES the need to see yourself in the blinding light of genetics come from?

Maybe it's because the biological view of humanity seems to be in the ascendant, pushing back decades of focus on culture. In the Western world, we have long convinced ourselves that everything human is just a social construction, that we have no *nature* but, on the contrary, are defined by our *culture*. And this seemed plausible, in particular, because most people live in a culture of which they could easily, painlessly, and pretty much automatically feel a part. A culture of which their ancestors have been a part for generations.

People were Danes, because they belonged to a special Danish culture – something having to do with pickled herring, Hans Christian Andersen, and red and white flags. South of the border, the Germans identified with Goethe and Schiller, schmaltzy music, and the Oktoberfest. Even in melting-pot nations such as the United States, people made reference to distinct cultures – black, Jewish, born-again, Latino, WASP, and countless others. Everyone could have a marvelously clear sense of his or her cultural affiliation.

Now, however, we are living in a world that is much more blended and stirred, like a well-mixed Martini, where even societies that have traditionally been homogenous are now describing themselves as multicultural. And as soon as you start to look a little more closely at the concept of culture, it is terribly difficult to define. A culture is a little like water – it has no fixed contours. The more the world is globalized and cultures are mixed, the more difficult it becomes to fasten down and, ultimately, to define your identity in this old-fashioned way. What culture do you belong to if your parents were born in Pakistan but you were born in Britain and have every intention of living the rest of your life in that country? Do you have more in common with Pakistanis your age or your British peers?

In this cultural cacophony, it is handy to be able to refer to something biological – for example, digital information that you can read and decipher and print out. I am a Zulu, or a Kpelle, because it is written into every one of my cells, no matter where I find myself or what culture I'm familiar with. Identity is in the genetic bar code.

WHAT ABOUT ME, Lone Frank? Can an ordinary "ethnic Dane" find some form of identity in her genetic background, and where do I even go to look?

There is, of course, the Genographic Project, founded in 2005, which in exchange for just a hundred dollars tests what they call *deep ancestry*. That is, where our early – very early – ancestors came from. Several years ago, the project met at an expensive restaurant in central London to hear the young American geneticist Spencer Wells explain how, using genetic analyses, he had traced the human migration out of Africa and around the globe. Over finger food and white wine, Wells described central Asia as "the playpen of humankind." Humanity's cradle was Africa, from which the first wave of modern *Homo sapiens* migrated and spread. It had long been believed that Europe was peopled by an emigration via the Middle East. However, when Wells and his colleagues analyzed DNA samples from population groups all over the world, they determined that the steppe peoples of central Asia had made the move west – not just to the rest of Asia, but to Europe, India, and the American continent, over the course of several population expansions.

Since the swanky London gathering to celebrate the publication of his book *The Journey of Man,* Wells has been promoted to the august position of "explorer-in-residence" at the National Geographic Society. He isn't resident much. Eternally wandering, like any good explorer, he serves as the photogenic front man for the Genographic Project, which is a joint undertaking between the society and IBM. More than anything, this enterprise appears to be a genetic search for our collective identity as a species. As Wells himself puts it, "In this future-obsessed era, it is important to seize a snapshot of our past before it is lost forever, in order better to understand ourselves and where we are headed." In his view, this snapshot is procured by collecting and comparing DNA from hundreds of thousands of individuals, who represent all the ethnic and tribal peoples of the world.

The project's ambition is to map in detail how different groups and peoples are related to each other and how they have moved around and mixed over the millennia. As reported in the *American Journal of Human Genetics,* the project's geneticists have uncovered that the past's great sailors, the Phoenicians, are the ancestors of the modern Maltese. Similarly, they have looked deeply into the genome of today's Lebanese population and analyzed how the migrants of the past left behind not only stones and monuments but chromosomes. A study of just under a thousand Muslims, Christians, and Druze revealed that, while the Christians largely bear traces of a European heritage, which presumably derives from the Crusaders, the Muslim population shares many genes with people from the Arabian peninsula, which can be attributed to the expansion of Islam between 600 and 700 CE. Curiously, they did not find any genetic traces from the Turkish Ottomans, who swamped the area in the sixteenth century.

More recently, the team has thrown itself into a new reading of our very early evolution, back when the first *Homo sapiens* lived exclusively in Africa and had not yet set eyes on the rest of the world. Paleontologists have discovered fossil remains of *Homo sapiens* that are two hundred thousand years old. Yet, despite the fact that three-quarters of the story of our development as modern human beings took place in Africa, that period has largely been ignored, in favor of untangling the events that occurred after the first migration out of Africa, around sixty thousand years ago. The Genographic

Project is changing that. Now, studies of DNA from a large group of living members of the Khoisan people – the Bushmen of southern Africa – are developing a new and interesting picture of the human past.

It seems there were very small groups of people, who dispersed into two distinct populations, which were separated by climate and a large desert that spread around today's Lake Nyasa, between Malawi, Mozambique, and Tanzania. For a hundred thousand years, these distinct populations, one in the northeast and one in the south, constituted their own evolutionary line. Genetic analyses indicate that the forebears of the Khoisan people may have comprised a distinct line, different from the others, for far longer than one hundred thousand years, and not until forty thousand years ago did they mix with other peoples, from the north.

MOST ANCESTRY STUDIES use the help of two keys: the male Y chromosome and *mitochondrial DNA*, a tiny circular chromosome found in small structures, mitochondria, which are the powerhouses of cells. Both types of chromosome have the special, and crucial, feature that they are passed on from generation to generation in virtually the same form. All the other chromosomes exchange genes left and right, meeting up in pairs and mixing freely when egg and sperm cells are made. This genetic dance is called *recombination*, and it mixes up the available genes into ever new combinations. Just like a deck of cards that must be shuffled before each deal, so each round of egg and sperm is not like the one before. This means, in practice, that the chromosomes you inherit from your father and your mother are not identical to the ones your parents carry around.

By contrast, the Y chromosome does not, for the most part, recombine, which means that a son's Y chromosome is a replica of his father's. Kinship testing consists of examining how the Y chromosome looks with respect to a band of special markers. The most frequently used markers are small "cassettes" of repetitions of short base sequences – for example, GCC – which are somewhat like genetic Lego blocks. The blocks are found in special positions on the Y chromosome and, at every position, you can count the number of repetitions and use this number as the relevant marker.

Women have no Y chromosome, but they are responsible for passing on mitochondrial DNA. This molecule is transmitted unchanged from mother to child – both to sons and daughters – because the mitochondria of embryos are provided exclusively by the egg. The sperm cell is such a scaled-down model that the only thing it brings to the party at the embryo's cell nucleus is the father's contribution of twenty-three chromosomes. The markers tested in mitochondria are of the good old SNP type, mutations that change one base for another and which can be found in particular regions of the mitochondrial DNA.

The markers we carry constitute our genealogical table. They are, so to speak, a catalog of the mutations that have occurred over the course of genealogical time, and by looking at the markers we can divide male and female descent into different lines. You can think of it as a tree. At the root is the clan father. Every time one of his descendants forms a mutation on the Y chromosome that is passed along to the next generation, the tree branches. The same holds true for clan mothers and mitochondrial DNA. Because geneticists know how often new mutations occur, it's possible to compare markers in two living people and calculate roughly how far back in time their last common clan father or clan mother lived.

For both women and men, researchers from the Genographic Project have been able to use the markers from thousands of people, living in different regions of the globe, to place humankind onto different branches of our family tree. In scientific jargon, these branches are called *haplogroups*, and the older they are, the more subgroups (or thinner twigs) they have. An illustration is the seven common haplogroups that are found in European women, each of which has developed and provided the source for many subgroups found in particular geographical regions.

Interesting differences can also be observed in a population's male and female descendants. For example, in Iceland today, Y chromosomes are primarily Nordic, while mitochondrial DNA is primarily Celtic. This is due, it appears, to the fact that Icelanders descend from Nordic men – the fierce Vikings who fled from Norway to conquer new lands – and the Celtic women snatched as these conquerors sailed past Ireland.

In Europe, you can see traces of the struggles between primordial hunter-gatherer groups and invading farmers. Recently, a research team reported that the most widespread haplogroup among European men – R1b1b2 – came from farmers who spread from Anatolia eight thousand years ago. These healthy swains penetrated the continent's original stock of men. Their Y chromosome constitutes eighteen percent of European men, and is known as haplogroup I. And those Anatolian farmers must have appreciated the local hunter-gatherer women, because today's European populations carry mitochondrial DNA that can be traced to these foremothers.

~~~

"HERE AT THE National Geographic we were all surprised that the interest in being tested is so great," says Spencer Wells, when I call him up at the Society's headquarters in Washington, DC. He is apparently taking a breather between two exotic expeditions and has a little time to spare.

"The guys in the marketing department estimated that we would, at most, be able to sell ten thousand test kits – if we were lucky. Because, honestly, how many people today can really be interested in knowing what route their Stone Age forefathers took to get to Asia or Europe?"

There were ten thousand who wanted to know exactly that on the very first day the test was available. Five years later, just under four hundred thousand test kits have been sold, of which eighty-five percent have gone to Americans. As Wells says, "Europeans may find this enthusiasm harder to understand, because they still have a connection to their villages."

I wonder just what he means by that.

"Over there, there is a sense that your kin have always lived in Denmark, in France, or wherever. There is an ethnic identity connected with being a European. That doesn't exist in the US, where all Americans are hyphenated – African-American, Cuban-American, Chinese-American or some mixture across ethnic demarcations."

On the other hand, this mixed America offers a bounty from the genetic ocean. Wells is working on a film project in which he visits a particularly multi-ethnic neighborhood in Queens and takes DNA samples from two hundred randomly selected inhabitants.

"We pretty much find the whole human variation in the form of haplogroups represented in a very small piece of American soil. And we can tell all of them who know nothing about their descent a fascinating story," he explains.

But familiarity with deep ancestry is not enough for everyone. I think again of A.M. Homes, who ordered a test from the Genographic Project and got the result "haplogroup U," which places her in the so-called Europa clan, which originates from a woman who lived about fifty-five thousand years ago and whose descendants have spread across the continent. "I feel I have used hundreds of dollars to find out something I already knew – namely, that I'm related to everybody else," noted the disappointed Homes in *The Mistress's Daughter*.

Spencer Wells thinks her attitude is regrettable.

"Of course, it is only a microscopic part of your overall descent, but it *is* a part of your descent," he stresses, a tad offended. "And most people get a kick out of it. To feel a connection to the first small groups of modern human beings in Africa and the whole fantastic journey the species has taken is something new and meaningful."

I myself can well understand how an identity-confused author would want knowledge that was closer to her: genealogy as opposed to anthropology. However, family tree research has also taken a step towards DNA. Whereas enthusiasts once pored over church records etched in Gothic letters and government censuses scratched in efficient shorthand, they can now test their DNA, and that of others, to get an unambiguous answer. With genetic genealogy, people can finally get all the way into the cells of the individual and see for themselves. Who is your real family and who is not? Papers can be falsified or misleading, but the signs written in DNA don't lie. If there is an "illegitimate" child somewhere, it can be covered with the right signatures in the right places, but the genome in the child and the child's descendants reveals what happened with unfailing certainty.

In the past few years, a veritable cottage industry has arisen in such genetic genealogy. In the United States alone, there are just short of fifty private companies offering genealogical tests of varying types and quality, and there are international genealogical organizations, interest groups, and networks.

Like my first confrontation with the hyper-extended family, this industry began with a question of Jewish identity. And it had to do with the group of Kohanim, who, according to myth, are descended directly from the biblical Aaron, the man who spoke to the assembly in place of his hot-tempered, stuttering little brother Moses. A Canadian kidney specialist, Karl Skorecki, belonged to this specific clan, and, when he ran into another Kohen in his local congregation, it struck him that they were very different in appearance. Skorecki was an Ashkenazi Jew, with fair skin and Eastern European roots, the other was an olive-skinned Sephardic Jew, with roots in the Spanish diaspora. But if, as the myth prescribed, these two men had a common forefather, it must be reflected in a common biology.

To test this, Skorecki went to the geneticist Michael Hammer. At that time, in the 1990s, Hammer was working with Y chromosomes in a lab at the University of Arizona, and took up the task with the curiosity of a researcher. Hammer did a detailed study of the Y chromosomes in Skorecki and a bunch of other Kohanim, and in 1997 published the results in *Nature*. The article became something of a sensation, for there was, in fact, a clear and close relationship, indicating a common descent in these men. Hammer found a characteristic pattern of markers on the Y chromosome that recurred in no less than 98.5 per cent of the men studied, all of whom had the surname Cohen. The pattern was promptly dubbed the *Cohen modal haplotype,* or simply CMH.

This was proof that the concept held water – you could use Y chromosomes to do genealogical analysis. But the leap from the academic arena to the free market did not come until 2000, when the US firm, FamilyTreeDNA, became the first to offer a commercial test of male descent. At around the same time, the British company, Oxford Ancestors, began touting a test of mitochondrial DNA. Soon, a veritable underbrush of genealogical concerns grew. By 2006, the turnover in businesses offering genealogical tests was sixty million dollars, and it is estimated that around a million people have already purchased a test, a number that is increasing by almost a hundred thousand individuals every year.

Genealogy is dependent on being able to compare gene sequences and markers, and this can happen either directly, between selected individuals who want their kinship tested, or by comparing a person to various databases. And the more data that are included in the database, the greater the chance there is that new clients will find unknown relatives and reconstruct a broader family history.

Amateur genealogists comprise an odd subculture, spending their holidays riffling through birth certificates and making rubbings of gravestones. Some might even call them fanatical. In the genetic age, there are hobbyists who take an obsession with their origins so far that they will dig into their own genome and that of others to map the past and to get an overview of the present. They learn to read DNA sequences the way the rest of us read the Sunday paper, and they are not shy about contacting complete strangers with the same last name and asking them for a bit of their DNA to check whether they are related.

If these strangers are reluctant, these genealogists have been known to go to great lengths to gather evidence. There are accounts of people merrily stealing DNA left behind by potential relatives and having it tested without their consent. Like the nice, elderly lady from Florida, who happily admitted to the *New York Times* that she tracked down a man she suspected of being a descendant of her great-great-great-grandfather's brother, in order to snatch a Styrofoam cup from which he had drunk coffee. Or the one who just as proudly bragged about how she pinched a hair – root and all – from a dead great-aunt who was on display at a funeral home.

"I UNDERSTAND THEM very well," says Bennett Greenspan, on the phone from Texas. The director of FamilyTreeDNA, the oldest and largest enterprise in the industry, admits that they do, from time to time, get illicit samples for analysis. With a nod and a wink, and for a little extra fee, they have extracted usable DNA from fingernail clippings, toothbrushes, and the backs of licked postage stamps.

"Why do people do this sort of thing?" he says, puzzled, when I ask. "Well, it's not hard to understand – they wonder where they come from and want certainty."

Greenspan knows, because he himself has roots in the amateur genealogical movement. In his sixties, with a bald pate and round glasses, he has managed to make his hobby into his livelihood. By 2009, the lab in Houston had conducted over half a million DNA tests, and the company's database now contains sequences from 176,000 Y chromosomes and mitochondrial DNA from 106,000 individuals.

"But we are always expanding the repertoire," Greenspan hurries to point out.

Thus, customers can submit orders to have their chromosomes tested far beyond the usual handful of markers. Among other things, FamilyTreeDNA has begun to sequence entire mitochondrial genomes, with all of their more than sixteen thousand bases, to go beyond the known mutations and, hopefully, discover some new ones. It is a way of getting closer to the present and capturing genetic events that might have happened a few generations ago in someone's great-grandmother. Mutations that only relatively few people share and which, therefore, are very specific.

"*That's* when it really becomes genealogy!" Greenspan says. Then, he shares the genesis story of his nice little business. In 1999, he lost his job, and didn't really know what he was going to do now he was unemployed. He had been dabbling for some time with a project to find related Greenspans, and map his family tree. His somewhat frustrated wife suggested that he do something serious about it and stay out of her kitchen. Bennett Greenspan took up the challenge but quickly ran into a brick wall. He found a man in Argentina who had the same last name as his cousin in the United States, but was unable to prove any kinship.

"The paper trail dried up," he recalls with a heavy sigh. New tools were needed and, because Greenspan had heard about Michael Hammer's academic studies of Y chromosomes among the Kohanim, he knocked on Hammer's door in Arizona. When he realized how efficient DNA was in tracing connections, it was clear that this was every genealogist's wet dream.

"Unfortunately, my own project is going rather poorly," he says today. He has tested dozens of Greenspans from home and abroad but has not found any he is related to.

"I'm like the shoemaker who has lots of holes in his own shoes."

"What about Alan Greenspan, the former chairman of the US Federal Reserve?" I ask, and receive an audible sigh.

"Oh, I don't know," he says with a slight pain in his voice. "I've tried, *really* tried, to get him to provide some DNA for a test, but he just won't do it. Maybe, that's for the best – he's a little too right-wing for my tastes."

He relates that, among those who look for genetic assistance from FamilyTreeDNA, Scots constitute by far the largest group.

"They are obsessed with learning what clan they belong to – it's very important to know whether you are a McDonnel or a McArthur or something else. Then come the Irish, the British, and the Ashkenazi Jews."

But, Greenspan assures me, he certainly gets inquiries from my fellow Scandinavians.

"If I have to explain why genetic genealogy is so interesting, it is because it provides a very personal experience of history and our connection backwards. There are stories and oral traditions in all families. Here in the US, it might be something about a great-great-grandmother who might have been a full-blood Cherokee or a great-grandfather who was black and, in Europe, it might be something about Jewish ancestors. As families, we all have our own creation myths, and a genetic test can ultimately confirm or destroy these myths."

Greenspan chuckles amiably into the phone.

"It's quite amusing," he says, "when I have to tell a man from Arizona who is Catholic and figures he is Spanish that he actually seems to come from Jewish ancestors."

The amusement is entirely on Greenspan's side, because I have a hard time imagining that situation.

"But it's not at all improbable, if you think about world history. The Spanish Inquisition chased a lot of Sephardic Jews out of Spain and forced others to convert to Catholicism. Some of the latter ones later ended up leaving the country, and their current descendants in the US, of course, consider themselves to be Hispanic."

"But do the people themselves also think it's funny to have their creation myths lying in ruins?"

"Not always. Some say 'well, that's that, then' and get on with their lives, while others take it very hard. Let me give you an example: A Jewish man who is deeply religious and believes he belongs among the Kohanim but finds out from us that he has the wrong haplotype. I've seen that sort of thing develop into a terrible identity crisis. There are also African Americans who get a shock when their Y chromosome tells them that they are Scots, because of a single white forefather. Everybody knows, of course, that black Americans mixed with whites, but to see it so directly in your genome is sometimes disturbing. Some have such a hard time with the results that they demand another test, because they are convinced we have made a mistake."

"And ...?"

"We never have."

But, in some way or other, an otherwise impeccable test *can* be considered a mistake. In the end, it provides a completely warped picture of reality, because it tests one line in a long series. If we go ten generations back, we have 1024 forefathers and foremothers, each of whom has, potentially, contributed equally to our total genome. Our chromosomes *are* a shuffled deck, bearing small calling cards from all our ancestors. But from the whole bunch, only a single man has donated the Y chromosome to be tested, and only a single woman has passed on her mitochondria with their circular DNA. What you test for your ancestry is your mother's mother's mother or your father's father's father and so on, while everyone else is eradicated.

"Yes," observes Greenspan, slightly troubled. "It's true that we come from many lines. I myself am a Newman or a Klein, if I focus on a different grandparent. People exaggerate the significance of one of the given surnames you come from, right? Of course, I am a mixture of many families, but my *identity* is certainly connected to the last name Greenspan."

Immediately, he is intrigued by my own surname.

"You think of yourself as a 'Frank,' right? Your identity is in some way wrapped up in your last name?"

When I think about it, it sort of is. My birth certificate reads Frank Pedersen, and Frank was originally just a middle name. But it was my mother's maiden name and when she divorced my father, taking me and my little brother with her, Dad's Pedersen was amputated without much ado. A pain-free little visit to the municipal courthouse. What did I think back then, at twelve years old? I can hardly remember. Not much more than that Pedersen was such a common name in Denmark, and wasn't it more sophisticated to be called Frank? In time, it has become an identity, because it is the by-line of all my articles, and the author's name on the cover of my books. I almost think of it as a graphic identity more than a familial one.

"You say that you're an 'ordinary' ethnic Dane, but don't you want to investigate whether there might be any trace of an Ashkenazi Jewish background? Frank, after all, sounds very Jewish, and I can help you."

That's very kind, I say.

"We now have 165,000 men in our database from 180 countries, and we will be able to see whether you are just western European or whether there might be some Ashkenazi ancestry mixed in. But you'll have to test your father."

I don't tell him that's impossible. Instead, I explain that my last name is my mother's.

"Aha. So, we have to get hold of a Y chromosome from your mother's side. Does she have brothers?"

My uncle is still alive, but I doubt it is possible to get him to provide a cheek swab.

"I haven't spoken to him in almost fifteen years," I explain.

"Oh, you can deal with that. Just go for it."

Greenspan, of course, is used to much pushier characters, but I promise to try.

"Excellent. Just let me know."

I WONDER IF I have what it takes to be a tough-as-nails family researcher. Could I survive in that world? Could I bring myself to procure a bit of "abandoned" DNA from my uncle – from the edge of a glass, maybe, or a cigarette butt?

As a first step, I pull myself together and call him. That is strange enough all by itself. After almost fifteen years – except for a two-minute phone conversation three years earlier – I don't really know what to say. But then he picks up the phone, and it seems as if we had just spoken together the previous week.

"Hello, little Lone!"

As if I were still the schoolgirl who came with my parents to visit at Christmas and a couple of times every summer. But he keeps tabs on things. He was always the one who was interested in the family, and he apparently enjoys the assignment. Holding the phone, he gets the family album down from a shelf. It lists the maiden names and birthdates of grandmothers and great-grandmothers – things I never knew anything about. While my uncle reads them aloud and makes comments at his end, I start sketching out a crude family tree at mine.

I know my parents' data, but it surprises me that my grand-mother on my mother's side was born in 1912 in Birkerød on the island of Zealand. I had always thought we were pure mainland Jutlanders, but apparently not. I have a few pictures of her mother, my great-grandmother, and I can remember my mother talking about her as "little Granny Hansen." She was a tiny woman, with round cheeks and hair parted immaculately down the middle. In old age, she lived with my grandparents, and within the family she was never called anything but her husband's surname.

"She was born in 1868 as Gjertrud Rosenlund," I hear my uncle say. *Rosenlund*. A very beautiful name. Not hard and brash like Frank, which has something unhelpfully masculine about it. Rosenlund – which means rose garden – is much softer, and, while I let the name roll over my tongue a few times, I think about Bennett Greenspan. When it comes right down to it, I could just as easily consider myself a Rosenlund. I carry just as many genes from my long dead great-grandmother as I do from my great-grandfather Frank, from the same generation, who just happened to give me my last name.

"But we don't have that name from him," I suddenly hear, though I don't understand. My maternal grandfather was named Frank, so his father must also have been named Frank? Right?

"No, his name was Sørensen. Hans Peter Sørensen, born in 1883 in Gjellerup."

The explanation comes from a law, passed after the World War II, which allowed people to take their mother's maiden name. My maternal grandfather did, because he wanted to open a bakery in a small town that already had a baker called Sørensen. So, the name Frank came to me through my great-grandmother, Ane Johanne Frank, born in 1886 in Silkeborg. A tall, desiccated woman, who let me ransack her purse for big, copper two-cent pieces, when I visited her in the 1960s.

"Interesting, huh?" my uncle says. So it is, but unfortunately it makes Bennett Greenspan's offer to search for any Ashkenazi descent impossible. For the Y chromosome that I would otherwise try to persuade my uncle to donate comes from great-grandpa Sørensen and does not belong to the original Frank family.

"You can also save yourself the trouble, because it's not a Jewish name," my uncle says. "If we go another generation back to Ane Johanne's father, he came – or so it appears – to Denmark as a 'potato German'."

For a moment or two, I can hear him leafing back and forth.

"Here he is: Johannes Michael Frank, married to Ane Marie Sørensen, who was born in 1852 at Sunds."

Potato German? There's not much glamour in that. I get the tale of a small group of German colonists who emigrated to Denmark around 1760, from the area around Pfalz and Hessen, to cultivate the heath of central Jutland. When the grain crop failed, they threw themselves into potatoes, introduced as a new and exciting crop to Denmark, and they became known as experts in cultivating the starchy, brown tubers. Thus, my roots were solidly in immigrants and farming stock. I shudder at the thought that it is not impossible, after all, that I am related in some convoluted way to the pale Eberhardt at the counter in the Frankfurt airport.

But what if my parents had never been divorced? Would I have experienced myself as a Pedersen and researched my father's side of the family?

So, I call my father's aunt, who is over eighty but sharp as ever, and the person who keeps tabs on that part of the family. Aunt Anna is the sister of my deceased paternal grandmother. She can, of course, rattle off when their parents, that is, yet another set of my great-grandparents, were born, and where it happened. They were both born Pedersens, so marriage changed nothing. It appears that Anna is also well informed about her sister's family-in-law — which also goes by the name Pedersen.

"Your paternal grandfather's father, Peter Pedersen from Skæring, him I liked," says Anna, sounding as though she means it. "There was something luminous about him."

On the other hand, his wife, Ane Katrine, whose maiden name and birthplace have been lost, no one really cared for. My father's account of his grandmother was no cozy bedtime story. This bony little woman was a slavedriver, who guided husband and six children with a hard hand. In her old age, she kept a son and a daughter at home on the farm, so they could take care of the business. When she died, at over ninety, the two children were so old that it no longer made sense to begin their own lives. They moved into a little house in one corner of the farm and lived out their last years there.

"Yes, they *were* a bit peculiar, those Skæring folk," affirms Anna. "And there was something about that southern blood of theirs. At any rate, people talked about one of the Spanish soldiers who were in the country during the Napoleonic Wars and went around burning and pillaging the south of Jutland."

Here, I begin to get a little interested. I myself have utterly ordinary dark-blonde hair and fair skin that, at most, acquires a yellowish hue in the summer. But there is something else in my father's family, if you look. My grandfather was very dark, almost olive-skinned, and had a nearly Roman hook to his nose. Of his three sons, one is fair-skinned and ruddy, while the other two are very dark. My father was raven-headed in his youth and his skin turned copper in the first rays of the spring sun. He could easily have passed for Middle Eastern or, at least Mediterranean.

You can test for that, I think, after hanging up. If my father was descended directly from this Spanish freebooter, it might appear in his Y chromosome, which is the same Y chromosome he passed on to my little brother. And since my little brother and I both received our mitochondrial DNA from my mother, I can make do with testing him to get all the pieces of my ancestry. I pick up the telephone again and, for once, get hold of him immediately. As a lawyer, he is not the brightest bulb as far as science goes, so I explain slowly and carefully what I want and why. At first, there is silence.

"A gene test?" he says suspiciously.

I assure him that it only involves a vanishingly small part of his total genome and that no one would be able to deduce anything about disease risks or anything else he wants to keep to himself. "No worries, it just involves some insignificant regions of your Y chromosome. And our shared mitochondrial DNA, of course."

He sounds as though he has tasted something abruptly unpleasant. "*We* share DNA?"

LUCKILY, BABY BROTHER is quick to get over his initial shock and turns up to deliver his DNA to me. Packed into his camel-hair Hugo Boss coat and clutching his briefcase tightly, he sits on the edge of an easy chair and rinses his mouth with a green liquid that catches loose cells from his mucous membrane. After the minute and a half the instructions prescribe, he spits it all back into the plastic cup from which it came. I say thank you and apologize for the taste, which truthfully is unpleasant. Then, I press down the lid and put the sample into a little cardboard box, together with the rudimentary family tree I had sketched.

The green cell solution has to journey a long way, to Utah, where it will be entered into a database. Since I couldn't make use of Bennett Greenspan's large collection of Ashkenazi Jewish men, I have decided instead to go to the Sorenson Molecular Genealogy Foundation, a non-commercial research foundation that collects DNA from around the world and is actively seeking participants from Denmark. The foundation also boasts of having the world's largest database of both genetic and genealogical data. There are 107,000 DNA samples collected from volunteer donors from 170 countries, each supplemented by the volunteer's family tree reaching back four generations, including name, birth date, and birthplace for everyone.

The project has a strange and fascinating history. Its founder was James LeVoy Sorenson, an eccentric who died in 2008, having amassed a fortune through real estate and a long list of medical inventions. Patents on disposable surgical masks and plastic catheters, among other things, catapulted him to forty-seventh place on *Forbes* magazine's list of the wealthiest Americans. Without giving up his business empire, Sorenson threw himself into genetic genealogy in the final years of his life.

"It began in the summer of 1999, when I got a call at two o'clock in the morning," explains Scott Woodward, who is today the head of the foundation. That summer, he was a professor in genetics at Brigham Young University in Salt Lake City, and had achieved some small fame in 1985, as the person who identified the first genetic marker for the lung disease cystic fibrosis.

"Anyway, the telephone rings at two o'clock in the morning," he repeats with his strikingly monotone voice. "And, as a father of four teenage boys, you take that kind of call."

Fortunately, it was neither the police nor the emergency ward but just an elderly man who, without much introduction, asked whether

Woodward knew anything about DNA. The professor thought he did. Fine, said the man, who then asked how much it would cost to "do Norway's genome."

"I was a little flabbergasted and asked what he meant by that, and then Sorenson finally introduced himself and explained that he was calling from Scandinavia. Apparently, he hadn't thought of the time difference."

Sorenson was on a tour to find traces of his Norwegian fore-bears and, since he was used to thinking big, he believed that he might just as well map the entire population of Norway while he was at it. "It sounded insane in the middle of the night, and I was not initially interested," Woodward says. "But he pressed his case after he came back from his trip and, in the course of few weeks, I could begin to see some interesting scientific questions that could be answered.

"For Sorenson, however, it was a very personal project. He was in his eighties and was searching for a connection between himself and his Norwegian ancestors. Perhaps, he even wanted to find living relatives in Norway."

I had heard of the Mormons' notion that you can save long-dead relatives from eternal damnation on Judgment Day by digging their identity out from the darkness of the past. Could that have been what Sorenson wanted?

"Of course, he was a Mormon, but I don't think it was religiously based," says Woodward, who becomes thoughtful. "But the project slowly grew larger. Sorenson wanted to collect DNA from every living Norwegian, and I figured it would run to half a billion dollars. 'You can't afford it,' I said, to which he answered: 'try me.' I did, and after a two-second pause, he said, 'half a billion, I can do that.'"

Woodward considered the matter and argued that you could use half a billion dollars better. "Forget Norway," was the message.

"Let's collect DNA from people from all over the world. If you get enough people in your database, it will be possible to take two people from just about anywhere and show how they are related and figure out when their last common ancestor lived."

I can hear Woodward take a deep breath on his mobile phone.

"The idea was that this knowledge would change how people look at each other."

Now it's my turn to be silent. Is he telling me that this is an idealistic project to create world peace by showing people that they are genetically related?

"You could say that," he says without noting my ironic edge. "We have collected blood and genealogical data for ten years now – our ambition is to be able to see kinships here and now and, at the same time, be capable of going, perhaps, five hundred years back in time. In addition – of the more than one hundred thousand DNA samples we have from one hundred and seventy countries, there are also some from Denmark, where I myself have roots."

That is all fine and good, but I am more interested in how you get all the volunteers for the project.

"We find them mostly by word of mouth. They may be people here in the US or in some other country who have heard about the project or know someone who has and then call to volunteer. We've had students organize the collection of samples. The condition for taking part is that the volunteers must be able to provide their family history four generations back, and that has proven to be difficult."

Most people, like me, could only account for three generations.

"That's why we have fewer samples than we planned – we could have had a million by now, if people knew their family history better. Oh, well, but as soon as we've collected the DNA and the information, our genealogists start searching the archives of different countries to trace people ten to eleven generations back."

I gasp audibly into the phone. Ten generations back means several hundred years of history and 1024 names. And you can multiply this by the over hundred thousand individuals in the database. Is this at all realistic?

"It *is* a huge job, but our little staff of twenty employees works like a mule," Woodward says contentedly. "But it's up and running now; you can go into the database and look. We have lots of success stories of people who have identified living relatives."

I've been in for a little peek. Sorenson has entered into a partnership with GeneTree, which is one of the many commercial companies through which anyone can buy a test and then compare him- or herself directly with the huge database.

There are quite a few accounts of stubborn amateur genealogists who eventually strike gold. Take the story of the woman from Seattle who found a perfect match for her mitochondrial DNA that led her to a family connection in Mali, which she could research in more detail. Or the incredible account of Mike Hunter's many-year effort to cast some light on the ancestry of his departed grandfather. In 1981, as Granddad Lindsey was laid into the ground, Hunter set out to confirm the family legend that Lindsey's mother, who had given him away for adoption as an infant, was a full-blood Indian. Armed with Granddad's birth certificate, Hunter followed a paper trail that led him to the backwoods of Virginia, and to a marriage registration between Nora and Albert Hunter who, it seems, were Lindsey's biological parents. The discovery was a bit of a disappointment, because both seemed to be very white, without a trace of Native American blood. Yet, it was a bit of a mystery that the two had not married until seven months after little Lindsey's birth, and there were some folks in the area who thought that Nora had probably been impregnated by someone other than Albert. But now it was 1985, and it was impossible to track any further answers ... the paper trail was gone.

Many years later, in 2008, Mike Hunter heard about the wonders of DNA and took up the case again. He volunteered for Sorenson's project and put his Y chromosome, which was also his Granddad Lindsey's, into the database, in the hope of finding some long-lost relatives. He looked, naturally enough, among the fourteen families registered under the name of Hunter but found nothing. After some time, however, there was an unexpected hit – a perfect match for the Y chromosome in a family by the name of Bailey, who came from the remote little town in Virginia where Lindsey's parents had married. Soon Mike Hunter was meeting his new family, the descendants of his real great-grandfather.

"This sort of thing is very gratifying," says Scott Woodward, a bit unctuously, "but, as I mentioned before, our foremost goal is to change people's view of each other."

This goal, he explains, spurred the foundation to go to Israel at one point and arrange a meeting between two women – an Israeli and a Palestinian – both average, ordinary people.

"I remember how the Palestinian, a little worried, whispered to me that she had never been so close to a Jew before," says Woodward.

I dread what is about to come.

"But we could show them that they share a lot of DNA sequences and were closely related as peoples. This broke the ice between them, and the story actually ends with them becoming personal friends."

Woodward is starting to sound like an evangelist, and I try to steer the conversation in a more technical direction. I wonder about the size of the future market for genetic genealogy and what the competitive parameters will be.

"A large database is obviously one parameter. But, in fact, I think the absolutely crucial thing will be the capacity and ability to interpret genetic information for customers. It is clear to me that personal genetic information will be an integrated part of our lives in many ways, but most people will have a hard time understanding and using this information. We need some interpreters."

I REALIZE WHAT Scott Woodward means as soon as I receive my own test report. It is filled with numbers and apparently random combinations of letters, and it makes me feel like I'm trying to decode a foreign language. Fortunately, I have an agreement with Ugo Perego, who heads up the laboratory in Salt Lake City. He has promised to give my chromosomes a thorough going-over. In the meantime, I find that they have tested my brother's Y chromosome for forty-three markers, each of which consists of a series of repetitions of short DNA sequences. The numbers I could not interpret at first are the number of repetitions of each of the forty-three positions. Unfortunately, the result is not particularly exciting.

"On your paternal side, you and your brother belong to haplogroup I1," says Perego in his thick Italian accent. This information does not tell me much except that my brother and I are Scandinavian. Haplogroup I1 pretty much only exists in Europe, where it seems to have arisen twenty-eight thousand years ago. Today, about one in three Danish men is I1.

"That's pretty boring," I let drop. "But what else could you expect?"

Perego declares his agreement. But then he adds, in a completely different tone: "The good news is that your mitochondrial DNA is definitely not boring."

Before sharing the results, he runs through an explanation of how mitochondrial DNA is handled during testing. In Sorensen's lab, the procedure is to sequence 1100 bases from the more than 16,000 of which make up the circular mitochondrial genome. These 1100 bases include some regions where mutations typically occur, and the sequence found in the test is compared to a reference sequence, the *Cambridge reference sequence*. That sequence was taken from a European woman who belongs to haplogroup H2. From the test, you can state which specific bases are different from the reference, that is, which SNP mutations the subject has compared to haplogroup H2. The more differences, the further you are from the standard, Perego patiently explains over the phone, giving as an example some African haplogroups, which typically exhibit fifteen to twenty differences.

I study my report and discover with a little disappointment that I'm pretty close to the reference. There are only six mutations given.

"Correct. But *what* mutations! You have a few that are old and quite rare, which means that we can place your haplogroup in a very specific context in relation to the history of human migration."

I wait while Perego rummages among some papers. Scientific literature, he says.

"First of all, you belong to haplogroup H2a1, which is a subgroup of haplogroup H2. Your subgroup has its origin in the Middle East and the Caucasus, where it arose about ten thousand years ago thanks to a new mutation. This is what we call 16354T in your report.

"A split in the line occurred in the Caucasus, where three branches of H2a1 emigrated, each in its own direction," he continues. "One branch went south to the Arabian peninsula, a second went to northern China, Russia, and Siberia, while a third ended up in Eastern Europe and, later, Western Europe and finally Scandinavia. In the branch that reached Scandinavia, a new mutation arose later – namely 16.193T – which you also have."

I check the report, and there it is.

"Would you like to have some numbers?" asks Perego, and I don't mind if I do. I have to admit he has done his homework.

"Today, H2a1 is found in four percent of the eastern Slavic peoples and about one percent of Estonians and Slovaks. There are even fewer among Scandinavians and, of them, it is far from everyone who has both the 16.354T and 16.193T mutations – the combination of the two, I would say, is very rare."

I admit I'm deeply interested. And it must be something I can use genealogically – I can take my rare mutations and hound down unknown relatives.

"Well, the names of your forefathers aren't imprinted in your DNA," Perego says, a little sharply. "If you want to answer genealogical questions, you have to have a theory about something that can then be illuminated by DNA. You're thinking of going to public databases and searching for a match, and this can be done, but the problem is that only a small number of those who have been tested are in the databases. So, it's a fluke if you find something. There are many more success stories in cases where two people who suspect they might be related are tested, so it can be confirmed or disconfirmed."

"But *if* I found a perfect match for my mitochondrial DNA," I persist, "then could you say something about how long ago our last common ancestor lived, or what?"

Ugo Perego lets out a long sigh.

"Because mitochondrial DNA mutates so slowly, a perfect match in itself is no guarantee that you have a common ancestor within the last four hundred years. But if you come from the same geographic region, it may be worth getting in contact through a database and investigating further. You would have to look at your own and the other person's genealogical data and, in this way, determine whether you could be related. That's how it works."

It is somewhat better with the Y chromosome, which mutates more quickly. You can figure that, if two people have a match of thirty-seven markers, there is a fifty percent chance that they share an ancestor within five generations.

"But it could also be seventeen generations, and suddenly it's very difficult, because you are back in historical times when there were no surnames," Perego says, sounding as if the conversation is now boring him a bit.

I think of the Spanish soldier who may or may not be in my family tree. If he is not a direct ancestor, even a thorough analysis of my brother's Y chromosome could not say anything about him. But I've heard there is something new on the way.

"Oh, you're thinking of autosomal analyses," says Perego, with a little more enthusiasm. Instead of looking at Y chromosomes and mitochondrial DNA, which only show one line of infinitely many, you will be able to see traces of many forefathers and foremothers by analyzing the other chromosomes. *Autosomes*, as they are called.

"We're working on it," says Perego, admitting that the new analyses really should already be accessible. "It has to do with being able to follow sections of chromosomes generations back and, in that way, get a picture of the combinations that occurred. But it is difficult to see which sections come from whom and, if you go farther back than four generations, it becomes a question of statistics. The information they can provide right now is, for example, that a person appears to have ancestry that is seventy percent Eastern European and thirty percent Scandinavian or something along those lines. We actually have a lot of data from our volunteers, but there are some technical challenges that have to do with statistics, and I don't want to bore you with them."

But there is something else as well, Perego adds. People are just not ready.

"What we are talking about here is far beyond the capacity of the general public. Generally speaking, most people don't even know how you search the accessible databases, because they know nothing about mitochondrial DNA or Y chromosomes. The result is that they try once, don't find anything, and then give up. It's a waste."

I can sense he is about to talk about the need for more interpretation and anticipate him.

"Exactly, exactly," he says, and he relates how the Sorenson foundation established a sort of emergency help line for the genetically challenged. You pay a symbolic sum and get a consultation with a specialist who can explain genetics back from Adam and Eve, so to speak, and who also helps with your searches. They even offer advice before people have been tested.

"These are people who are interested but who don't know what the different tests can even tell them and they can learn who in the family they should test to get a given piece of information. If I were to be completely honest, genetic genealogy today is a product aimed at the élite, the relatively few who have read up on things and are already well into the basics."

I'm not sure what group I belong to, but I would like to take up the hunt.

"Your 16.193T mutation is rare and is not found in everyone in haplogroup H2a1, so that's good to search," he offers as a tip. "And remember – the databases are growing all the time, so if you don't find anything today, it may be you will be lucky in six months or a year. Keep on trying. Good luck."

I'll need it. When I go to Sorenson's own database and search among Y chromosomes, I find no complete match, and there is no hit until I lower my requirements a tad. I ask the database for people who share eight out of ten genetic markers; it serves up a handful. But when I go in and look more closely at these examples, all of which have surnames that are anything but Pedersen, the statistical programs estimate that there are between nineteen and thirty-two generations to our most recent common ancestor. That doesn't help. I keep working but there is nothing in either the Y-search or the Y-base databases, so I quickly abandon the male line and shift over to search mitochondrial DNA.

This is not without hope. I know that one of the daughters of my great-grandmother Gjertrud Rosenlund emigrated to America in the previous century, and it is not entirely improbable that some of her descendants may have been gene-tested. Why not? Genealogy, after all, is the most popular hobby in the United States.

So I feel a little tingly when I start up the Sorenson search machine, but the result comes up lightning-quick – and empty. No one provides a complete match. I try again, searching by haplogroup, and it appears that, out of eighty thousand in the database, of which by far the most are of European extraction like me, only 385 belong to H2a1. Of them, a paltry seventeen have both of the important mutations, 16.354T and 16.193T.

For all intents and purposes, that's practically no one. And not a single one of them shares my other mutations.

I feel at once pleasantly rare but also unsatisfied, and I have a strong urge to keep hunting. The next possibility is Mitosearch, a database available to the public, from which you can gain access to individuals whose mitochondrial DNA has been tested by Bennett Greenspan and the Genographic Project. I follow Ugo Perego's recommendation and search my mutations. In fact, I find two that match. Both are men – one from the Ukraine and the other from Finland. Unfortunately, I can see that they have been tested for fewer markers than me, so it is impossible to know whether they also fit with my other mutations. It looks like the trail has petered out, but not quite.

By chance, I run into a volunteer DNA project, Danish Demes. It is one of the many examples of amateur genealogists who throw themselves into doing something that approaches research or, at least, lends itself to research: a group that tries to map the network of kinships that exists for Danes both in Denmark and around the world. At the moment, they have fewer than a hundred participants, but I spot at once that two belong to haplogroup H2a1. I check their mutations. They are not exactly like mine but not far off, and I feel so encouraged that I write to the project's anonymous administrator and ask to be let in from the cold.

"You would be very welcome," writes the organizer, who turns out to be an elderly, hard-of-hearing lady from Florida. A retired zoologist, Diana Gale Matthiesen is a genealogical veteran, with decades of experience and a clear veneration for DNA tests, which she calls "the best thing that's happened since the Internet." Her dear, departed father was Danish and, after researching her mother's family to the bottom and identifying many distant and unknown relatives, she wanted to get "closer to her Danish roots."

But Diana's ambition goes much further. She wants to construct a specialized database of Danish haplogroups and a catalog of the specific mutations found in individuals in each group. She envisions a project that will become a gold standard for people who believe they have Danish roots but are not sure where to go or with whom to compare themselves. At the same time, it will be a resource that true professional geneticists, anthropologists, and historians can use freely in their research.

From the outside, it seems bizarre. I've never given a thought to the distant past or unknown ancestors, and now I volunteer to provide a set of mutations to cyberspace. Propped by coffee and candy, I sit up into the wee hours of the morning, searching for strange people who share something as abstract as the chemical building block thymine in a particular position in the circular chromosome of their mitochondria. I'm searching for a trace of a small and, presumably, completely meaningless change that arose some thousands of years ago in a single woman. A nameless ür-mother who cannot be traced but has passed on a hidden sign to an unknown number of descendants. I wonder whether I will feel any kind of kinship if I found others out there in her multibranched family tree.

At any rate, the fascination has worked its way into me, and it is beginning to resemble a form of identification. Sometimes, I think of myself as belonging to a special haplogroup and the carrier of a special mutation and, in a way, these thoughts give me a strong experience of the very meaning of genetics. I can sit quietly with a powerful feeling that we are each of us temporary depositories of information that has an almost eternal life, and which is passed on and on and on. At the same time, the idea is a gateway to a genuine sense of broader kinship. A kinship that we, of course, know exists between us and the rest of humanity and the living organisms of the world, but which otherwise seems so abstract.

The whole experience is also a gateway to a fascination with the information itself. I want to get closer to my genome – my digital me – and decide to take the next logical step. I will put mitochondria and Y chromosomes behind me and take a completely different type of test, one that involves my entire genome. A test that casts a net over the six billion base pairs and captures some of the countless tiny variations it contains. Variations that, according to researchers, will provide a picture of my prospects for health and for disease – a picture not of my past, but of my future.

3

Honoring my snips,
in sickness and in health

DEAR LONE FRANK:

We are pleased to inform you that the results of your deCODEme
gene scan are now available. Please click here.

IT'S ALWAYS NICE to get a message that is straightforward, and I do as I'm told. A little tap on the mouse, and I gain access to my personal deCODEme page, which assures me that the company has tested my genome for more than a million genetic markers. Excellent. Some of the research has already found a risk of disease, and, as it says on the screen:

> *Your results include forty-six medical conditions. Find your risk and read about prevention.*

The message seems friendly enough, but my mouth still goes a little dry. Three weeks ago, at home on the sofa, it was pure fun and high jinks to rip open the envelope of instructions I had received through the post; nor was it any big deal to swab the inside of my cheek with the stick supplied by the kit and send my little scraping of mucous membrane back to Reykjavik for analysis. But right now, here in front of the computer, my desire to see into the future has abated. Somewhat.

The thing is that there is no such thing as a flawless genome. We are all mutants with ticking time bombs hidden inside. I know that theoretically, but as long as I don't have a precise picture of where those mutations exactly are, it all seems safely hypothetical, unreal.

To top it off, I'm alone and far from home. I've gone to Reykjavik to talk to the people at deCODE Genetics and have installed myself at a hotel on the city's main street. Although it's the start of spring and the sun is out, there is a thin but pleasingly decorative sheet of snow over everything, and the temperature has stayed stubbornly below freezing point. Iceland's capital is bleak and windy and, even in the hotel room, I'm constantly cold.

It's late in the afternoon and dark outside, and I have opened a large can of beer for a little company. I glance down at the "medical conditions," which are broken down into categories according to which organ they affect. Blood diseases, joint and muscle disorders, digestive problems, eyes, lungs, and throat. Cancer, brain, nerves. Hair, skin, and nails are there, too.

It's an indiscriminate mixture, as the eye wanders down the list: glaucoma, sclerosis, kidney cancer, asthma, gallstones, baldness (for men only), type 1 diabetes and type 2, restless leg syndrome, gout.

I pause. Gout? Is that even genetic? I thought that was something crusty men got after a lifetime of overindulgence. Too much foie gras and port, deposited as uric acid in the joints of the big toe.

IT IS SURPRISINGLY simple to put together a gene profile. Most of the work is automated and takes place inside machines and computers, almost untouched by human hands. The tiny amount of DNA in my mouth swab is cultured in a copy machine designed for that purpose — an apparatus that starts a *polymerase chain reaction*, in which specialized enzymes meticulously pull apart and multiply the DNA molecules, hundreds of times over. After this step, the whole thing is poured into a gene chip that measures a million tiny SNPs distributed throughout my chromosomes.

A very few of these SNPs are known to be connected to certain diseases, in that they increase or decrease the risk of getting them. Take, for example, someone who has acquired the unromantically named rs9642880, which is found on chromosome 8. By comparing SNPs in almost two thousand Icelandic and Dutch patients with bladder cancer with those in thirty-four thousand healthy "control" people, the researchers at deCODEme and a number of European universities discovered that rs9642880 apparently increases the risk of that cancer. One in five people of European descent has base T at this position, and it gives them a 0.5 times greater chance of developing bladder cancer at some point in their life than a person with base A, G, or C.

If you want, you can read through corresponding studies for other diseases and their markers. Likewise, you can inform yourself about the technical details that form the basis for the risk calculations. I go to the page with these long equations and have to admit that I don't much feel like dealing with them right now – I want to look more closely at my genetic horoscope, my digital self.

Where do you begin? I decide to start offensively, draw the mouse a little uncertainly over a number of cancers, and finally click on chronic lymphatic leukemia. The text presents the symptoms of the disease dispassionately: extreme exhaustion, headaches, bleeding, and depressed immune reactions. Finally, it explains that it only strikes half a percent of the population and, typically, after they reach seventy. Before I can get my personal odds, I have to click that I understand and accept: *Do I really want to see the results and do I understand what they represent?*

This is what the modern version of informed consent looks like. Some brief sentences on a screen and two clicks of a mouse. Of course, there is no doctor or researcher present who could explain to me what I might not understand, so what can I do? How can I stop myself from clicking the buttons, one after the other, without a second thought? It's a little like the fine print in contracts and insurance agreements – you just don't feel like reading it. But I know I'm capable of understanding a risk presented as a percentage, so I quickly click "Accept."

Ha! It begins well. My lifetime risk for chronic lymphocytic leukemia is only 0.2 per cent, two out of a thousand. That is almost forty percent less than the average for white European women. I feel as though I've just won on a lotto scratchcard.

I look down at the other choices and happen on intracranial aneurysm – a scary out-pouching of a major blood vessel in the brain that, without warning, can burst and leave you a vegetable.

Another winner! Only a three percent lifetime risk, which again, is half the normal rate. Now I've tasted blood, so I check age-related macular degeneration, or AMD, the progressive breakdown of the retina that was the subject of one of the first association studies and is the most frequent cause of diminished vision and blindness among older adults. Researchers have found gene variants in five regions of chromosomes 1, 6, 10, and 19 that come into play and affect the risk. I think of an acquaintance back home in Copenhagen, a woman who has been an avid reader and a professional translator of literature all her life but is now pretty much blind, relegated to audiobooks and dependent on the help of other people to get around. I hesitate for a moment. My fingers hover over the mouse. Yet once again I click, and find myself far below the average population risk. A paltry two percent, versus the normal eight. I toast myself in the mirror, which happens to be hanging over the desk, and decide to take on one of the heavies, while my mood is buoyant.

Alzheimer's. The most frequent cause of dementia, it says. Here, I can't help but think of old James Watson and his fear of carrying the ApoE4 gene variant, with its significantly increased risk. But when I read my results, they are outright fantastic – not a single ApoE4 variant in sight. Instead, I have two copies of ApoE3, which is connected with a low risk of Alzheimer's disease and gives me just a seven percent chance of developing the devastating disease, whereas the population average is twelve percent.

My dad was right, I *do* have good genes! And spontaneously, I think of a friend at home and how she keeps urging me to get down to business and propagate before it's too late. "When you're hale and hearty, highly educated, and fairly bright, it is a *duty* to society to have children who can raise the standard," she lectures. My friend is herself highly educated, extremely bright, and has so far done her duty to society three times.

But it's time to stop stalling. I know where my problems are likely to hide – we all do. I have a family history of diseases running through generations. On my father's side, it is cardiovascular problems. My paternal grandmother, my grandfather, and my father – all weak hearts and lousy blood vessels. My grandmother suffered from an indeterminate "bad heart," as it was called then. My grandfather had poor circulation and sores on his shin that wouldn't heal and, even though he worked like a horse all his life and was apparently in fine form even at a high age, he was struck by a debilitating cerebral hemorrhage right after hitting seventy. And my father. In his forties, he was diagnosed with seriously high blood pressure – so high that the family doctor called for an ambulance to freight him directly to the nearest hospital. Somewhat later came the trouble with his heart, which fluttered and would not beat regularly. Pills and medication piled up and were changed endlessly, but they could not prevent him from developing arteriosclerosis in both legs. In the end, there was no pulse left. My father, who had built his own house and did handsprings in the back yard, could barely walk and silently fumed about his physical decline.

"I'll do fine as long as I've got the radio and a library full of books," he said, but, after a while even sitting in a chair began to hurt so much that he just wanted to get away. It was peripheral arterial disease, or PAD as it is now popularly called. And sure enough, deCODEme had PAD listed, right under the category of heart and circulation. They can inform me that the condition strikes more than one in ten people in the industrialized world and that smoking is the grand villain behind it. But genetic factors have also turned up; Icelandic researchers have discovered a single meaningful variant on chromosome 15. I take a deep breath and click.

I knew it – it doesn't look good. Now, my personal risk is suddenly increased in relation to the norm, to around eighteen percent as opposed to fourteen. It may not be a colossal difference but, in absolute terms, almost one in five. As if that weren't enough, deCODEme even stresses that their test "may not include all risk factors."

I no longer smoke but get clammy palms thinking about my younger years, when I was a party smoker. Weren't there an awful lot of parties back then? I check for what they have to say under prevention and treatment, and it shows that you can take blood-thinning aspirin to reduce the risk of PAD. Maybe I should take it up with my doctor, when I get home.

It has made me tense and, since the can of beer is long gone, the only way to curb the feeling is by attacking the minibar's store of chips and chocolate. In the mirror, I catch a glimpse of myself flaying the wrapper off a Snickers bar and think: "Come on. Get it over with."

Heart disease and hardening of the arteries are not nice options, but my greatest anxiety is reserved for something else. What I'm really afraid of is breast cancer. A horrific disease that, somewhere in the back of my mind, I'm certain I'm going to die of, just like my mother and my maternal grandmother.

Some of my earliest memories are of visits to my grandmother at the radiation ward of Århus Hospital. Grey linoleum floor, high glass doors, and weak Kool-aid. When it got to be too much with the adults, who were all sitting very still, I could play with the little boy with a large brain tumor. They had shaved off all his hair and indicated with a purple marker the areas of his head that were to receive radiation treatment. It looked like some exotic map, and I remember those purple lines almost better than my grandmother's illness – but then, I was only four years old and did not understand the gravity of the situation.

It was infinitely worse when I started secondary school, and it was my mother's turn. One day, she came home from work and told me and my little brother, with whom she lived alone, that she had cancer and was having her left breast removed the next week and that the doctors didn't know how bad it was.

There were tears, angst, and hope that it had not spread, but it *had* – possibly not very much, so there was hope that radiation and chemotherapy would work. They did, and my mother was free of disease. But only for six months. Then, there were spots on the scans. Small spots but still . . . It turned into more than three years of hope, examinations, disappointments, new treatments, agony, depression and, eventually, pain. Nights when she screamed and whimpered because the pain from the bone metastases was difficult to curb even with morphine.

Some might see it as a kind of magical thinking that I stay far away from any actual examination. I never fondle myself to search for lumps and, when anyone asks whether I shouldn't have regular mammography, I usually mumble some defensive witticism as "they can't do that for such small breasts."

Now I'm sitting in front of a bright but cold screen that says that it can reveal the risk carried in my genome as defined by eight variations on chromosomes 2, 5, 8, 10, 11, 15, and 16. For women of Caucasian descent, the average risk of developing breast cancer is around twelve percent. Twelve out of a hundred ordinary women will get it at some point – no way around it. But depending on the combination of these seven variants, you can come up into the vicinity of a sixty percent risk. I feel certain that I am one of the six, and keep my eyes closed long after I've clicked.

But it's unbelievable: I have a lower than average risk – not even the ordinary twelve percent, only 7.7. It is as if a very old, hissing pressure deep inside my body quickly seeps out and floats away.

For a wonderful, absurd second, I believe I've been liberated from death.

"WOW. YOU LOOK like you could use a strong cup of coffee."

It's seven o'clock the next morning, when deCODEme's head of public relations, Edward Farmer, comes to retrieve me from the hotel. It's not just cold in Iceland, it's also dark, and to boot, I haven't slept well. First, I had to celebrate my great relief at my lowered risk of cancer with a couple more beers, and later, when the euphoria wore off, I conscientiously studied my whole gene profile, one end to the other. From atrial fibrillation and baldness to gout and restless legs syndrome – all delightfully low. Sure, baldness was a given, but I wasn't going to deny myself the delight.

Still, my night's rest was uneasy. I wound up wondering how usable these genetic risk assessments really are, when it comes right down to it. Not just for me, but generally. What do you do with the information that you have a slightly smaller risk of getting asthma than the average? And what does it mean for your everyday life that your risk for kidney stones is a bit higher? What would I have done if my gene profile had given me a sixty percent chance of getting breast cancer? And how sure are they, anyway, about these molecular risk assessments?

"Put your seat belt on," says Farmer, "I'm sure Kári will be able to explain all that."

By Kári he means Kári Stefánsson, deCODEme's founder, director, public face, and a bit of a story unto himself. The last time I had the pleasure was over ten years ago, when deCODE Genetics – the parent of deCODEme – was an infant and a hot topic of debate. The company was among the first to latch on to industrial genetic development, which was just beginning to shake things up. Stefánsson had a good idea, and he planned to realize it by drilling into the genome of the Icelandic people. He imagined a gigantic database in which the medical records of living Icelanders would be collected and combined with their genetic information and the genealogical records of the population going back centuries. *Here* was an efficient way to trace the genes that make us sick.

Stefánsson scraped together twelve million dollars from a group of believing venture capitalists, left his pleasant and prestigious position as a professor of neurology and neuropathology at Harvard University, and went home to Iceland, where he began to steamroll his will through the legislature. There was bitter local opposition from those who thought the genomes of a population should not be put into the hands of a private business, and the outrage spread far beyond the borders of this tiny island nation. Stefánsson became a sort of itinerant verbal prizefighter, famous in the media for his predilection for dark Armani suits and for his wrathful, Viking-like appearance. He went from conference to conference to address his critics and quickly gained notoriety for his intrinsic stubbornness.

Now in his sixties and majestic at six feet tall, with broad shoulders, white hair, and a beard, Stefánsson is, for good measure, the descendant of Iceland's beloved poet and national bard of the tenth century, Egill Skallagrímsson. Ironically, Egill was not only known for his unusually beautiful verse but also for his unusually horrible behavior. If the food and drink were not to his like, the great poet might vomit on his host and, if anyone offended him, he did not shrink from gouging out their eyes or maiming them in some other way. Today, some of the labels tacked onto his descendant are "boastful," "aggressive," and "notoriously gruff."

On deCODEme's homepage, however, Stefánsson is portrayed with a warm smile, his chin resting thoughtfully in his hands. A winsome grandfather, were it not for the tight, black T-shirt that emphasizes his buff upper arms, and the text beneath the portrait:

> *Unlike a large number of other companies hoping to follow on our heels, deCODEme is not just a website that happens to be about human genetics, it is your portal into the world's largest and most*

> *successful effort to understand the inherited risk of the most common*
> *diseases in contemporary society.*

Arrogantly put, but essentially correct.

Stefánsson's concept has been proved to hold water. For ten years now, deCODE Genetics has ransacked thousands of Icelandic genomes, from which scientists have identified an impressive battery of gene variants that affect human health and the risk of disease. It regularly spits out articles for the best scientific journals. Farmer has been kind enough to supply me with a pile of the most recent.

Professionals are impressed by the research but not everyone thinks it is a good idea to offer gene profiles directly to consumers, as the company now does. The skeptics include, among others, the UK Human Genetics Commission, which is worried that ordinary people cannot really understand the information without thorough, professional, genetic advice. The commission's members argue that tests simply stating a risk should not be sold directly to consumers. Many doctors are of the same opinion and, in 2008, pressure from physician interest groups prodded the State of California to prohibit businesses from performing genetic tests that were not ordered by a licensed medical professional.

That seems to be the trend. In 2009, after years of economic difficulties, deCODE Genetics did a U-turn and attempted to reinvent itself as a "diagnostic company." It filed for bankruptcy and was saved at the last minute by a group of American investors. Soon there will be money – lots of money – to be earned on gene profiles. As Stefánsson says in the press release that announced the decision: "As the focus of our healthcare system shifts toward prevention, measuring and controlling the individual risk of disease will become a central part of everyday medicine."

IT HAS BEEN ten years since my last meeting with Stefánsson, but his greeting tells me that he has not changed much.

"Are you really going to burden me with your company once again?" he says, extending his hand without rising from his chair or looking up from his papers. Before I can answer, he shouts to his assistant in reception that he needs some coffee if he is going to survive this chat. A classic tack from his repertoire; one a number of my journalist colleagues have written about. What surprises me is not his rudeness but his voice – I had completely forgotten how pleasant it is. Soft, as if wrapped in fur. His English has a thick accent that is really quite charming, even now, as he tells Farmer that as head of communications he can be present during the interview but should be prepared to keep his mouth shut.

"*Completely* shut."

I figure that the only way to avoid being completely shut out is to make the first strike. I begin by poking an old wound. Stefánsson's start-up was the first to market the personal gene profile, but its competitor, 23andMe, based in Silicon Valley, is far more well-known, and presumably more commercially successful. I ask with an air of innocence what Stefánsson thinks about the fact that 23andMe featured prominently when *Time* magazine crowned the gene profile as "invention of the year." A brief but very cool look greets me from the other side of the desk.

"It did not surprise me, not at all, and I'll tell you why. One of its two founders shares a bed with a rich, famous man, and that's the sort of thing Americans fall for," he growls. "Okay, I was pissed off for an afternoon. But so what? I've been pissed off a lot of afternoons in my life." He looks directly at Farmer, who commendably maintains his PR mask of neutrality.

I recall that Anne Wojcicki of 23andMe – the woman who "shares a bed" with Google's founder Sergey Brin – made a very successful appearance, while visibly pregnant, on the *Oprah Winfrey Show*. Somewhat later Stefánsson was left to explain the blessings of gene profiles to the lower-rated Martha Stewart, the ever-smiling goddess of domesticity whose stint in prison for securities fraud was then still recent news.

"Tell me ... who is the typical customer for deCODEme? An average viewer of the *Martha Stewart Show*?" I ask, hoping to sound innocent.

"Where the hell did you find this one?" says Stefánsson, apparently to Farmer, but this time his growl is accompanied by something resembling the early stages of a smile. The ice is broken, and we can finally start the conversation in earnest. The great man shifts in his chair, turns for the first time in my direction, and explains that the company sells most of its gene profiles to professionals.

"General practitioners, but also clinics that specialize in prevention, which are shooting up all over the US," he explains. "The number of private customers is increasing, of course, but in the long term the healthcare system is the biggest market. In fact, I consider our direct marketing to consumers as a way of influencing the healthcare system. If you believe that prevention is ever going to amount to anything, you have to have far better information about the individual's risk of disease. This is the key to limiting the burden of disease."

The question is whether gene profiles are the future. In January 2008, a group of American geneticists warned, in an editorial in the *New England Journal of Medicine*, that there isn't much clinical value in using them. The major illnesses that strike the general population – diabetes, cardiovascular disease, cancer – are complex. A series of genes is involved, but insufficiently understood. That's true also of the multitude of environmental factors in play. It is uncertain whether gene profiles can provide a realistic picture of the risks that an individual faces.

"That's pure nonsense!"

Now a tenor of anger has entered into his voice, and Stefánsson abruptly leans forward.

"A gene profile is as useful as any other form of screening. What would you do if it shows that you have a fifty percent chance of developing breast cancer? The statistics say that the disease is ninety-nine percent curable if it is diagnosed early, but is almost never cured if it is only discovered after it has spread. I think the percentage of the population that is cured would look far better if genetic profiles were used more diligently."

The example gives me the cold sweats, but then I remember my fortunate gene variants and calm down. In the meantime, Stefánsson continues his soliloquy.

"We could also take cardiovascular disease. Our test has some markers for cardiovascular disease that provide a better prediction for the risk of heart attack than traditional measurements of cholesterol in the blood. For people who believe in preventative treatment, gene profiles are useful. But, of course, only if they react on the basis of the information."

Even when he is speaking, he is restless in a way that is reminiscent of a hyperactive child. He constantly fiddles with his mobile phone and regularly demands more coffee – without at any point offering some to Farmer or to me. I hope to capture his attention with a point that I know will irritate him.

"We already know what we need to do to minimize our risk for the major prevalent diseases," I say.

We should quit smoking, stuff ourselves with vegetables, make sure to exercise, and keep our weight down. But this knowledge does not necessarily make people change their lives, so why should it be any different if they know some numbers on their genome? For most people, chromosomes, markers, statistics – they're all an abstraction.

Slowly, Stefánsson squints his eyes and considers me, the way you'd consider a colorful but disgusting insect.

"You talk about 'people' as though they were some sort of inferior race, and then you babble nonsense on top of it." He taps his fingertips on the table to emphasize his words. "The experience from cholesterol measurements is that people take it quite seriously. About thirty percent of those who are found to have high cholesterol follow up by taking medication to lower it and changing their lifestyle. Another thirty percent react, but a little more half-heartedly."

Of course. And forty percent ignore the information. In addition, cholesterol levels only say something about a single disease, whereas a gene profile provides risk estimates for several dozen different illnesses. Who can rationally take a position on that?

"You totally misunderstand. When you look at a gene profile, it is extremely rare to see an increased risk for more than one serious disease. I'm acquainted with one individual, whose profile I've seen, who has an increased risk of two major diseases. It is simply incorrect that you place an enormous burden on people by having them consider a heap of risk factors. And another thing is that when you have risk profiles for a significant part of a population you can as a society begin to shape targeted preventative health plans. This will probably be the only navigable way to limit costs in the future."

I ask whether they should do gene profiles on all of us at birth. Stefánsson thinks that's a very good idea. The more information there is, the greater the positive effect it can have for the individual over the course of his or her life.

"Listen," he says, sounding tired of the discussion, "a lot of people wind up misunderstanding and misinterpreting this data. But since all life on this planet is based on DNA sequences and research is supplying an ever-deeper understanding of how these sequences affect the individual, it is inevitable that we will come to use it. It would be *criminal* to decide that people must not use this knowledge before we know exactly how it will affect them biologically, psychologically, and socially. You can consider what has been going on for the past few years as the clinical trial phase that takes place after a drug has come on the market. And, in this case, it is even driven by the consumers themselves."

Now he apparently cannot sit still any longer, because out of the blue he asks whether Farmer has seen his latest series of pictures. Without waiting for an answer, he asks us to follow him. We trail him to a room down the corridor, which is utterly empty except for eight huge photostats leaning against the walls. They are extreme close-ups, taken on a beach not far from deCODEme's offices, of pebbles and wet seaweed, blown up to a meter and a half and printed on aluminum plates.

Stefánsson describes the special lens they were taken with and talks about how much he loves to traipse about in nature with his camera as his only companion. I ask a little pointedly whether ordinary rockweed could be so brilliantly purple or whether his exposure has gone wrong. "That's how it looks in reality," he replies, lecturing me on the wondrous and unique Icelandic landscape.

"It's the most beautiful place in the world."

His pictures are quite good, and this irritates me for some reason. It doesn't seem fair that you can be a model physical specimen, reach the top of the academic world, establish and run a high-tech business, and be a sensitive artist. In the back of my head, I once again hear my father's speech about "good genes."

How does some packaged, inactive genetic information specify that we become the people we are? Among these six billion base pairs, only a few percent separate a randomly chosen human being from an equally randomly chosen chimpanzee. What separates one human being from another is a mere half of one percent. How, I wonder, can these relatively few chemical changes have such wide-ranging significance and provide such different results as the kind, courteous, and curly-haired Farmer, the dauntless, sharp-tongued, black-clad Stefánsson, and me?

"Well," he says, interrupting my train of thought. "You have had a gene scan done yourself, I understand. Is there anything in it you want to ask about?"

BACK IN THE office, this time without a chaperone, I confide my test results and complain about my higher risk for arteriosclerosis. Since my first perusal the day before, I've read that the tested variant also increases my risk for lung cancer. It is up to thirty-three percent, which sounds quite overwhelming to my ears, but Stefánsson is not impressed.

"It doesn't mean a thing, if you don't smoke. Even though you are genetically predisposed, your risk falls to almost zero if you don't smoke."

I promise to stop but then say that I'm also nervous about my increased risk for basal cell carcinoma, the most common form of skin cancer. Here, too, there is no cause for special alarm.

"I've had one removed, it doesn't mean all that much. You should just stay out of the sun."

I already do but confess that there is something I don't understand. I've been thin and bony my whole life, but now deCODEme reports that I have a higher than average risk of being overweight. Me, fat? I ask Stefánsson whether there could be a mistake in his test.

"No," he replies with a conspicuous chill in his voice. "I myself have the same obesity variant as you but have weighed the same for many years – maybe because I exercise so much. Preferably, three hours a day. Otherwise I get depressed." It may be the thought of the workout room, lined with dumbbells to beat off the blues, but he gets up and grabs a bottle of water from a small refrigerator in the corner of the office, and even offers one to me.

"Variants of multifactorial diseases will never give you a precise prediction, because they are complex and are not determined by a single gene. The outcome is powerfully influenced by environment. You have some gene variants but their *penetrance*; that is, to what degree they affect an individual, is a mystery. We can't say what they do, whether some variants will be expressed or others will not."

I gawp at him. Is he actually admitting that gene profiles can't really be used for anything?

"No, I'm not. A gene profile can tell you that your probability for a given outcome – being overweight, for example – is greater or lesser but it cannot predict whether you will actually become fat."

To illustrate the point about environmental factors, Stefánsson points to a finding in deCODE Genetics's research. The company scientists have studied melanoma and found a variant in the gene for the melanocortin receptor that behaves very differently in different European populations. For Spaniards, the variant triples the risk of melanoma, while the same variant plays no role whatsoever among Icelanders – presumably, because the sun also plays a role. In Iceland, you can avoid sunlight without any problem, which is somewhat more difficult to do so in Spain. In Sweden, where there is more sun than in Iceland but less than in Spain, the variant in question provides a risk that falls somewhere between the other two populations.

"We have also discovered three gene variants that increase the risk of atrial fibrillation. Each of them is three times more frequent in China than in Europe, but the Chinese suffer atrial fibrillation far more rarely. That is more difficult to explain than about the sun and melanoma. And it shows that we have to understand the influence of environmental factors far better, before we can use gene profiles with great precision."

Before I can object again, Stefánsson proceeds to yet another factor that complicates the picture. Apparently, it is not insignificant whether we inherit a gene variant from one or the other side of the family, that is, whether the gene in its time was provided by the egg, or by the sperm cell that won the fertilization race. Researchers at deCODEme could glimpse this mysterious effect, because they not only have genetic information from almost forty thousand Icelanders, but also data about their common relatives. Using a special type of analysis, they are able to see, according to Stefánsson, "one example after another" of variants that affect us differently according to which parent delivered it. For instance, deCODEme found that one such variant increases the risk of diabetes by thirty percent if it comes from your father, but decreases the risk by ten percent if it comes from your mother. These opposing effects were such that no one had noted the variant before, because in the classic studies the statistics cancelled each other out. In the *Nature* article reporting the discovery, the researchers were also able to identify a variant that increased the risk of skin cancer but only if it comes from the father's side. The same effect holds true for a variant associated with breast cancer.

Stefánsson is now ready to move on to missing heritability. "The variants we have found so far do not explain a particularly large part of the heritability of different diseases," he says.

I try not to roll my eyes. Yes, thanks, people do talk about the "dark matter" of the genome.

"You probably also know that many of them think it's about us going out and finding a lot of rare variants," he adds in a tone that makes it clear this idea is idiotic.

I mention that James Watson talks about rare variants, but I am ignored.

"*I* think it will turn out that a significant part of what we can't explain today will be able to be explained by effects that are dependent on which parent the ordinary variants come from. On the whole, I think it will turn out that our models for inheritance are too primitive and must be changed."

Here, he's touched on something interesting. Textbooks that may have to be rewritten, the possibility of truly novel discoveries. But what about gene profiles – doesn't this mean that quite ordinary people who were not born in Iceland cannot get a precise risk assessment? Because they don't know whether a given variant comes from their father's or their mother's side?

Unfortunately, when I take a breath to ask more, a secretary pops her head in and reminds the boss that he's scheduled to leave. He is on his way to an incredibly important event in California and the plane won't wait. So, to say at least one positive thing, I hurry to tell him that I'm happy about my low risk of breast cancer. I describe, relieved as I am, how my grandmother died before she was sixty and my mother at just forty-six, but I can quickly tell from Stefánsson's face that I'm celebrating too early.

"The common variants which we test are typically not involved in what we call familial cases. With that family history, you should get a specific test of your BRCA genes."

This prods at some knowledge that I've repressed into a distant corner of my brain. Mutations on the genes BRCA1 and BRCA2 are found in two to five percent of all women, and they provide an extremely high risk for breast cancer. Upwards of eighty percent, according to some studies. Maybe I should just stick with the gene profile.

"And hope for the best? I can't advise that," says Stefánsson.

But I can't see what good would come from the test. If I have BRCA mutations, I wouldn't be able to do anything but ruin my life with chronic worry and spend all my time scurrying through the doors of the healthcare system with pleas for mammograms and ultrasounds.

"That's not true," he replies. "You can have a bilateral mastectomy done."

Bilateral mastectomy. In Latin, it sounds so neutral and harmless. But is this man really sitting there telling me with cool objectivity that I should just have both my breasts removed? Get rid of it all, in the name of prevention?

"Yes, unless you consider that a fate worse than death."

⸺⸺⸺

I DON'T KNOW if I do. But I can't get it out of my mind for the rest of the day. Of course, I know that many women, as soon as they get a positive BRCA test, have their breasts removed and replaced by silicone. Whereas this radical solution was considered extreme just five to ten years ago, it has become almost mainstream in the United States.

Back at the hotel, I go immediately to the computer and find Myriad Genetics, the American company that has a patent on a diagnostic test of BRCA genes. They provide a quick questionnaire that is supposed to tell me whether the test can be used. They ask whether anyone in the family has had breast cancer before they were fifty, to which I answer yes. Then, I learn that I could easily have an increased risk for both breast cancer and ovarian cancer. And if I were an Ashkenazi Jew this risk would be even further increased, Myriad kindly explains. As I now know, I'm not, but the company's BRCA analysis can clarify the issue for me, and as they emphasize: "Understanding your risk of cancer is the first step toward managing it."

You can move on to a table and calculate your risk for having BRCA mutations, and with only one close relation who developed breast cancer before fifty, I'm down to 4.5 per cent. That is, just under a five percent risk of having gene variants that provide between a sixty-five percent and eighty percent risk of developing breast cancer. From which, of course, there is a varying risk of dying – depending on when the cancer is discovered and how it is treated.

The question is whether just under five percent is so small that I should simply pretend it's not there – according to the homepage, there is, after all, a ninety-five percent chance that my BRCA genes are in great shape. And if they are, because I am lucky to be free of certain other variants, I have just under an eight percent risk of developing breast cancer, according to my gene profile from deCODEme. On the other hand, wouldn't it be better to get rid of the uncertainty about whether the worst could happen once and for all?

Just one more test.

I'll have to think about it.

I STEP BACK and examine the genomic experience inside out and upside down. What do you get from a gene profile? If you were to look at it critically, you could say that, for all its more or less transparent risk calculations, it does nothing but remind you that you really shouldn't think you are healthy just because you think things are going well. In reality, "healthy" people are just patients who haven't yet been diagnosed. And once this acknowledgment has been solidly planted in your mind, you just wait around to get sick, right?

There is a minor army of ethicists and social scientists with an interest in medical technology who, from time to time, air the view that it is impossible for ordinary people to benefit from having access to these new genetic tests. Ordinary people just can't relate to the percentages with which Kári Stefánsson and his people fill their heads, and the information makes them nervous. The consequence is that they will be constantly running to their doctors, burdening an already overburdened system. They will live a more worried, stressful, poorer, and, perhaps, shorter life.

Many opinion makers and commentators are also unconvinced by the new testing opportunities. They include the British journalist Camilla Long of the *Sunday Times*, who penned a long, passionate tract in favor of prohibiting consumers from having direct access to genetic services. Referring to the percentages for risk of various diseases, she asks, "... who – in their right mind – would want to know? ... The impact on one's life of that kind of Damoclean diagnosis is almost impossible to imagine." Her gloomy conclusion for society: "The cost to our collective mental health is incalculable."

It sounds logical enough that people would feel burdened, but I don't think it holds water. At any rate, *I* don't feel that way. With my trip to Reykjavik at a proper distance, I can see that my deCODEme results have not emboldened my inner hypochondriac. I don't sit around brooding over the possibility of having my legs amputated because of PAD or, for that matter, of developing glaucoma and going blind. The risk is there, of course – and it is even relatively high – but it would be there regardless of whether I had been handed a number for it or walked around in my pre-swab ignorance. Admittedly, the thing about BRCA is lurking and nagging deep inside me, but no more than it did before.

On the whole, my degree of nervousness and anxiety about disease and death has not changed in any notable way. There is no unambiguous answer as to whether genetic predictions will make people nervous or not. Some will take it calmly, while others will freak out at a slightly increased risk of cardiovascular disease. And they will be the same people who, with the knowledge we already have, adopt a calm, wait-and-see attitude or become wild hypochondriacs. The reaction has to do with the individual's temperament and attitude rather than the information itself

So, the big question is whether I will change anything, do anything differently in my life, now that I have this specific, tangible genetic knowledge. Do I see myself in a new way? Have I managed to rewrite my future?

Perhaps, but not in the work-out-three-hours-a-day or remove-my-breasts mode that Stefánsson suggested. Instead, in a sense, I feel more present in my biology. Or maybe I should rather say I view myself to a higher degree as an *organism*.

I know it sounds strange. Especially because the concept of an organism brings to mind a being low on the evolutionary scale — microbes, worms, that sort of thing. Human beings are people or individuals, not organisms. But it is as if I have been given an extra insight into myself, an X-ray vision that does not see the *person* but rather a well-organized, teeming whole of perpetually communicating cells.

The idea of being an organism is surprisingly agreeable, almost cheering. It's as though the idea itself removes the burden of being an individual, because you can see yourself from a much grander perspective. My gene profile has given me an insight into some of my organism's strengths and weaknesses, but also, at the same time, knowledge of the buttons that I can regulate.

Just a moment, the critics will say: We all know that we can and should keep the old carcass in better shape; we can all rattle off the standard health advice in our sleep. What's new about that?

Conceded. We know it intellectually — but the *feeling* that you are an organism and that you are dynamic and malleable is much stronger and comes through more powerfully as a *desire* to shape yourself. A desire to do something. It is not just that, knowing that I have more than a thirty percent chance of developing glaucoma, I have made an appointment with my eye doctor to have the pressure in my eyeballs checked. Rather, I'm not going to drag myself to the fitness center under a heavy cloud of sluggishness. Nor do the weights and machines there continue to make me tired; now, they positively exhilarate me. While running on the treadmill, I imagine the complicated biochemical processes going on in my leg muscles, how all these splendid chemicals are being pumped into my bloodstream, and how they'll eventually get to my brain, where they'll influence my mood, my thoughts, and, essentially, my outlook on the world.

My personal deCODEme gene profile has opened a tiny window into this physical inner self, and it is like peering into a new dimension: a strange inner universe of digital information that unfolds and is transformed into physical manifestations we can sense and measure. A tiny piece of information comes into play in certain cells in the eye and, suddenly, the pressure in the eye is increased and your vision becomes poorer. Another genetic slip-up makes beta cells in the pancreas become lazy and to produce too little insulin – with diabetes as the consequence. I dig into my raw data, wanting more. In the special genome browser that deCODEme makes available, I leaf at random through my million SNPs and get used to their strange names: the prefix *rs*, a number that states its position on a given chromosome, and then the two bases you have been given at that position from each parent: rs4610 (T,C). But, honestly, how can we only have found fifty diseases out of a million genetic markers? There must be more.

AFTER A FEW late evenings spent this way, I find something that can give me a fix. Promethease, it is called. The work of two idealistic Americans, Michael Cariaso and Greg Lennon, Promethease is a free computer program that can wring information out of the deCODEme data by linking individual SNPs with the scientific studies that have investigated their meaning. Cariaso and Lennon wrote their code in 2007, and put it on the Internet for general use. There it stays, developing alongside the research, because Promethease has a built-in function that constantly monitors the major scientific databases for new articles dealing with SNPs. Every time a new connection between gene variants and traits is published, the program incorporates it into the collection and users get it with their reports.

"We get at least ten thousand hits a month," Cariaso tells me via a Skype connection from Amsterdam. "It's not hard to squeeze out massive quantities of genetic data, but to get them interpreted and into the hands of the individual customer is a completely different question. In the future, it will be a crucial competitive parameter for companies that want to sell genetic analyses."

That sounds very likely. At any rate, I took my hit, pulling my raw data down from deCODEme's server and sending it to a server in the United States, where Promethease is hosted, and I've now got a report safe and sound on my hard disk. Instead of the fifty or so conditions deCODEme decided to tell me about, I can now see over four thousand specific SNPs listed. It's enough to give me sweaty palms. The many variants are arranged into a handful of boxes marked with titles such as "Most Interesting SNPs," "Medicines," "Medical Conditions," and "SNPs Most Unique to You." There is also a catch-all box: "Even More Complicated."

Cariaso has generously offered to help me examine my report. "I can quickly see if there is anything interesting, and I promise to get rid of the file immediately afterward," he assures me.

I'm not shy about my genes, however. In fact, I have no qualms at all about having a complete stranger look at them on his computer without being there myself. It's one of the paradoxes of the genome: on the one hand, my genes constitute all the essential information there is about me, but on the other, they are at such an abstract level that they don't seem very private at all.

"Hey, Lone! You are the first rs8177374 (T,T) I've ever seen!" Cariaso writes quickly via e-mail. I almost blush. It appears that two percent of the Caucasian population is fortunate enough to have T,T in this position. To judge from the literature, that T,T is very good to have.

"A beneficial mutation that provides resistance to several diseases, such as invasive pneumococcal disease, bacteremia, malaria, and tuberculosis," Promethease supplies.

Very good.

"Beyond that, your report is quite unremarkable," says Cariaso. "But remember that boring genes are good genes. If you have an interesting report, it means trouble."

Excellent. I throw myself into the Promethease interface. First, I tackle the category for SNPs that have something to do with medical conditions, and I'm quickly attracted by rs2217262. It plays a role in the development of autism, it appears. Almost nine out of ten Caucasians have two As at this position in the gene DOCK4, while the rest have a C or an A or two Cs. I myself belong to the last group and, according to one study, we have less than half the general risk for autism. As it says in the article in Wikipedia to which you can click through directly from Promethease: "This is a protective gene variant."

Now, autism is nothing to worry about as a forty-three-year-old, but if I were to have children, my protective DOCK4 variant might have a soothing effect. Lord knows, it's an academic discussion, but there is more immediate relevance in the fact that I have a variant that seems to protect against "age-related mental decay," namely, the base T in my SNP rs3758391. It makes sense that this would offer protection, because it has to do with an old acquaintance: this SNP is found in the middle of a gene named SIRT1, which has long been a focus for aging research. The gene appears in similar versions throughout the animal kingdom. Even yeast cells have a corresponding gene, which if it mutates give the puny creatures a longer life.

In human beings, , for example, a study of a thousand sturdy Finns over eighty-five years old shows that the bearers of at least one

T were cognitively better off and even showed a slight tendency toward better heart health than did other people. I'm a little annoyed that I also have a single C – I would rather have been among the rarer, and presumably better protected, T,T genotype – and wonder where the C comes from. I only remember one case of dementia in the family. My maternal grandfather's mother, whom everyone just called "granny," survived in fine physical shape well into her nineties, but was lost to dementia in her last years. In the halls of the old people's home, she waited for the bus and couldn't understand where her little boy had gone. It was no use telling her that he had retired long ago.

"Hey, what about this?"

My boyfriend sneaks in to look at the report, and his near-sighted eyes have, of course, alighted on rs2146323. "It produces a small hippocampus," he says cheerfully. "That can't be good. As far as I remember, this brain area plays a role in learning and memory."

I confirm that the small sausage-shaped structure plays a crucial and central role for both.

"You have the genotype C,C," he says, reading aloud from the study. "Individuals with the C,C genotype display a hippocampus of less volume than bearers of T and A variants."

He looks at me triumphantly. But I remind him that, just a couple of years ago, I had my brain scanned by some top neuroscientists at the University of California, Los Angeles, and none of them said anything about a small hippocampus.

"That's lucky," he says, but there's more. "Increased risk of depression with this one – rs3761418, which is on the BCR gene on chromosome 22. Here, too, you have the G,G genotype, and they have a third greater risk for depression than carriers of the other variants."

Before he really gets going, I click on the original publication and point out that the result comes from a study of 329 Japanese patients. "Japanese!" I say with emphasis, explaining that you need to take all this with a grain of salt. It is not necessarily the case that each of the genetic associations that are discovered and written about holds up in practice. For a finding to have weight, it must be repeated in several independent studies, and it is absolutely not the case that you can automatically infer that a result that applies to one ethnic group also applies to another.

"I see," he says slowly. "I just think this variant is interesting in consideration of the fact that you *do* suffer from depression, right?"

He has a point. At the moment, I'm not medicated but, according to my psychiatrist, I ought to be – for purely preventative purposes. In the past seven years, I've had three depressions requiring treatment, and that is enough for them to know that, statistically, there will be more. They also know that every single episode harms the brain, and, quite apropos, one of the side effects of repeated depressions is that the hippocampus shrinks.

I've thought a great deal about my depressive tendency, because it clearly runs in the family. But I don't have the energy to go into it right now, and certainly not with an uncomprehending boyfriend reading over my shoulder. Instead, I would rather look at something I know Promethease focuses on – namely, how the body metabolizes various chemicals and, especially, pharmaceutical drugs.

As it happens, specific gene variants can tell you what dose you should take for optimal effect and, at the same time, avoid toxicity and side effects. As Cariaso put it: "It is so much more important to know something about how the body metabolizes drugs than about whether you're predisposed to Alzheimer's, which you can do nothing about anyway. I predict that genetic tests for how any number of drugs are metabolized will soon be a part of our medical records."

This would not be a bad idea. It is known that one in five deaths in the United States is due to faulty medication, and a good number of these are to do with doses that are too high for patients who metabolize the relevant chemical too quickly and are thus poisoned. Seen in this light, I am amazed how little deCODEme has to say about this sort of thing. The deCODEme report says I need more than 2.5 milligrams a day of the blood thinner Warfarin if I'm hit by a brain thrombosis. It also informs me that I am among those who, for genetic reasons, do not have a high risk of developing muscular dystrophy if I take cholesterol-lowering statins. Nice to know, but it doesn't mean much in relation to the fact that, on the basis of the same raw data, Promethease refers me to articles regarding the metabolism of more than fifty named drugs.

Methadone, for example. In a single SNP in the gene for one of the brain's special transport proteins, I have a T from both my parents, and this means I am among the three percent of the population who need more than 150 milligrams of methadone a day to counteract withdrawal symptoms from heroin. As my boyfriend notes, "It's almost to the point that you should keep a card in your wallet."

I'm not concerned about ever having to deal with heroin withdrawal or methadone doses, but it is worrying that it doesn't look like I can metabolize the drug modafinil very effectively. The drug was developed to treat the sleep disorder narcolepsy, but has also been shown to be excellent at keeping healthy people awake far beyond their normal bedtime while, at the same time, increasing concentration. In fact, students and shift workers who need an acute increase in late-night cognitive ability already use it as a sort of mental doping, and I have to admit that I have more than once thought of trying it. But, according to Promethease, a single unfortunate SNP means that I would probably experience no effect. All I can do is take comfort from the fact that a quarter of the population is in the same boat.

"This is really interesting!" my boyfriend now exclaims. "Listen – there is a gene that causes small breasts if you drink too much coffee."

At first, I think he is pulling my leg, but it's true. The CYP1A2 gene codes for an enzyme that determines how quickly we metabolize a number of compounds, one of which is caffeine. In rs762551, there is either a C or an A, and an A creates a change in the enzyme that allows you to metabolize caffeine more quickly. I myself am an A, C and thus have an overall metabolism somewhere in the middle.

"One study of healthy, fertile women concluded that those who drank tea or several cups of coffee a day had significantly lower breast volume. But only if they had at least one C in rs762551," my boyfriend recites aloud.

I have no desire to see his smarmy grin.

"Over forty percent of women of European extraction have this mutation," he informs me. "They should prohibit women under twenty-five from drinking coffee. It's clearly very deleterious."

I ignore the tasteless remark and read on myself. The researchers behind the breast volume study cite other reports indicating that coffee actually protects against breast cancer. A study of 411 women with mutations in the BRCA1 gene (of which 170 were ill and 241 healthy) focused on the connections between breast cancer, coffee consumption before the age of thirty-five, and the CYP1A2 genotype. And even though the CYP1A2 genotype does not in itself have anything to do with the risk for breast cancer, it appeared that women with at least one C variant in rs762551 who also drank coffee had a sixty-four percent lower risk than those who never drank it. Coffee had no effect on those women without C variants.

"Okay," I say, pulling the plug on Promethease for the day. "I'm going to keep drinking as much coffee as I usually do, but you know what? I think I'm ready to have a BRCA test."

GETTING SCREENED FOR BRCA requires more effort than sending a gob of spit through the mail. First, I have to go to my own doctor and convince her that I should be referred for genetic counseling.

"It's probably reasonable enough," she says, glancing over her reading glasses. She signs the form and sends the referral to the Copenhagen University Hospital.

Several weeks later, I receive a notice to turn up for a consultation – with a nurse, it turns out. A nurse who must not express herself on whether I can get a gene test or not, because her job is to find out if I even have a reason to pester the specialists. Do I have a sufficient aggregate number of cases of breast cancer in my family or am I just overexcited? She questions me patiently and takes notes on my family history. Then, she sends me home and begins to gather the available medical records about the sick and the dead. If we are lucky, she explains, there may still be a little tissue in alcohol or paraffin left from one of them, tissue that can be requisitioned from the hospital's biobanks and tested. While this is investigated, I'm supposed to wait – for several weeks – for an appointment with the clinical geneticist, who will provide counseling, if I'm approved for it.

"It is expected that the consultation will last about an hour and, therefore, a parking pass will not be issued," reads my referral notice. I bike to the hospital and find my way to the sixth floor and the department of clinical genetics. Behind the glass doors, there are no patients to be seen, only a couple of laboratories and a number of offices. My consultation is supposed to take place in one that does not call much attention to itself – discreet grey walls, the usual indeterminate institutional furniture, a noticeboard with a couple of wry children's drawings. An experienced clinical geneticist, Dr. Kjergaard, informs me that she has advised hundreds of patients. She opens a folder with my information, puts on a professional yet cheerful tone. She seems to be in a good mood, generally.

"You've been referred by your doctor, I see. I've got the relevant papers on family matters, and we have drawn up a family tree. If you look here, men are indicated with squares and women with circles."

It is an incredibly simple family tree for an almost pitifully small family. I am indicated at the bottom as a blank, white circle, but the two circles directly above it and connected to my own each have a large, fat line through them and a little black spot. Signs that stand for illness and death. Below that are written the dates of death, and for a moment I stare silently at the 1984 that is my mother's.

"I presume you yourself are completely healthy?"

How do you answer a question like that? It's impossible to know, and thanks to my genome, I've had my eyes opened to the fact that I am encumbered with a risk for one thing or another or another.

"So far as I know," I finally say, whereupon Kjergaard points to my family tree.

"We have documented information that your mother got breast cancer at forty-three."

"I've also just turned forty-three," I say without really wanting to. The remark hangs awkwardly in the air.

"Your grandmother got breast cancer at fifty-seven, I can see, and your grandfather died of prostate cancer, but that is not unusual at that relatively high age. Do you know of any other instances, on either your father's or your mother's side?"

"No. There's not that many of us."

"The exercise here is about assessing whether the aggregate number of cancer cases are more than you can expect in an average Danish family."

Kjergaard cocks her head and reminds me gently that cancer is a very frequent illness, which a third of us can expect to get at some point in our lives. No rarity, then.

"And breast cancer is the most ordinary form in women; every ninth or tenth is hit. But what can be said in your instance is that there are two direct, first-degree relatives who have had it. But to be able to say with certainty that it is something predominantly genetic and not just chance, we have to be able to see three generations, and we have to have more instances of cancer in each generation."

It shakes me a bit that two cases are not enough, but I can't come up with more. My maternal grandmother's sister emigrated to the United States when she was quite young, and I have no idea what happened to her or her descendants.

"As things stand, I can't say with certainty that there is anything hereditary at play here. But it stands out that your mother got it at such an early age."

See, that's what I think.

"What we do then is compare statistical analyses of families in which we've seen something similar and, in that case, you come out a little higher than the average. You fall into the category we call 'moderate lifetime risk'."

Moderate lifetime risk — what does that mean? It doesn't immediately tell me much. In most contexts, moderate is a positive word.

"I normally don't care for putting percentages on things, because it can be too much for some people. I'd rather put it this way: you have a far greater chance of getting through life *without* breast cancer than of getting the disease."

That sounds a little evasive, I think.

"But if you would really like a number?"

I would and, at last, a piece of paper with some data slides across the table. Kjergaard shows me a simple chart marked by two curves, one of which represents the average cancer risk for women between the ages of forty-three and eighty-three and the other my personal risk during the same years. My curve rises somewhat faster than the average and, at eighty-three, it reaches twenty-three percent.

"Your lifetime risk, which we calculate on the basis of some studies and tables, lies somewhere between twenty-three percent and twenty-eight percent and, since you thereby are a little higher than the average woman, the system has a particular offer for you."

A little higher? The phrase ricochets around the room. I know the average woman has a risk of ten percent. Further, I know we are looking at something more than a "little," and assume that we must be getting to the BRCA test. I move to roll up my sleeve so they can take a blood sample.

"Our offer is that, from now on, you can get an annual checkup here at the University Hospital, where you'll get a breast examination from an experienced breast surgeon, and a mammogram. For young women — uh, like you — an ultrasound examination is also performed."

What is she talking about? An experienced breast surgeon?

Here I am, sitting in the very heart of the Danish healthcare system, across the desk from one of its most prominent specialists, and their offer is to have me fondled and illuminated just like in the old days? I'm a member of the genetic age; I want a molecular insight into my situation, one appropriate to the times. It may well be that they can't see three generations back into my family, but that doesn't mean that they can preclude a BRCA mutation that may have arisen in my grandmother or even my mother. And which, therefore, I might have inherited.

Presumably, the experienced Dr. Kjergaard has encountered this type of mutinous attitude before, because she remains tranquil.

"If your mother had lived, we would have offered to do a BRCA examination first of her and, if she had mutations, then of you. But we normally only make that offer when there is someone living who has the disease, in order to ascertain what mutation it is. Since we are not sure whether there even is any dominant heredity factor in your family, the probability of finding a BRCA mutation in you is not particularly great. Where we typically find mutations, the family history looks somewhat different."

I feel neither comforted nor convinced. Why not just sequence both my genes – BRCA1 and BRCA2 – and compare them, base by base, with the mutations already known that can be taken from various databases? After all, we are not talking about smashing atoms but about tossing a little DNA in a sequence machine and getting a computer to read the printout.

In fact, I've come prepared. I went to the Myriad Genetics homepage, which explains that, in cases like mine when you don't already have information about particular genetic variations, the scientists there do a full sequencing and compare the client's genes with a consensus sequence for both BRCA1 and BRCA2. If they find deviations from the norm, they can go to databases and see whether they have been described before and whether they seem to increase the risk of cancer. At the same time, they look for types of mutations that you can reckon are harmful, because they prevent a protein from being produced by the BRCA gene. The healthcare system should be able to do the same thing, I suggest, and place my printout on the table in front of my advisor.

She looks a little tense and pokes at it with one finger, then takes a deep breath before answering my question. "Our attitude is that this requires counseling. People cannot themselves interpret something like this," she maintains. "Technically, we could easily do what you're asking, but we would still have an interpretation problem. Our procedure is to test someone who has the disease. It is the only way we can be sure of finding a connection between the disease and any mutations. But listen: even in families where there is a large number of cancer cases and we are certain there is a dominant hereditary factor for breast and ovarian cancer, we find a BRCA mutation in under a third of those who have the disease. This indicates that there are still risk genes and mechanisms we don't know about yet. The problem with full sequencing is that you find so much you cannot definitely know the meaning of. "

"Are you saying that, if they did a full sequencing of my BRCA genes, there could be changes that you haven't seen before?"

"There is a reasonable chance of finding something we cannot advise on, because we don't know whether it leads to an increased cancer risk. And that message would not be much fun for you to get, right?"

No, it wouldn't be, but neither is it fun to walk around with the feeling that there might be something that could have been caught and interpreted. I decide to press one more time.

"If I were very nervous and unconvinced by your explanation, and if I pestered you to have a full BRCA sequencing to see whether there might be something that they know about from other places …?"

The good doctor looks frazzled.

"Then, I would reply that the probability of finding anything is very small. Your family does not seem particularly at risk, and I've never had anyone ask in this way. But, as a starting point, if I can sense that you understand what the limitations are, I would discuss it with my colleagues."

I concentrate on getting my puppy-dog look right. And it apparently works, because Dr. Anne-Marie Gerdes, the newly instituted head of cancer genetics, is sent for, and she agrees to hear the case and help me make a decision. She especially wants to hear whether I've really understood the sort of a snowball I will have started rolling if I succeed in getting a BRCA test and it actually shows known mutations – mutations known to provide a sixty-five to eighty percent chance of breast cancer. Could I handle it?

I think so. After all, I have lived with these thoughts since I was around fifteen years old. I realize what choices I will have if I discover I have any mutations. I can accept the offer Kjergaard already proposed – annual palpations and mammograms – or I can have both breasts removed and replaced by silicone implants, as Kári Stefánsson so kindly proposed. I ask whether they have any guidelines for what they are recommending.

"We don't talk about that sort of thing," says the chief surgeon, who gets an almost indignant look in her eyes. "This is genetic counseling, and my job is to make sure that the person I'm counseling understands the options and chooses what is right for them. I have no agenda or position on what they should do. All that about having the breasts removed – and the press *love* to write about that – very few end up choosing that. It's a big decision and something that people resort to in families in which a slew of young women have died and in which, after conversations here and with breast surgeons and plastic surgeons, a woman figures out that this is what she wants. Then we support the decision."

"It is a procedure that takes more than a full year," adds Gerdes in a steady voice. "First, they remove both breasts and nipples and place a tissue expander under the breast muscle. They pump a little salt water into them at the beginning and, over a few months, they are gradually expanded. Then, you have silicone implants put in and a reconstruction of the nipple done with tattooing."

"Yes, it's a question of whether it's worth all that trouble at my age," I say, hoping to get another reassurance that I'm still young. It doesn't come, but the chief surgeon suddenly broadens the conversation to include a new organ.

"We also have to remember that there are ovaries to take into consideration. Mutations in BRCA1 especially increase the risk of ovarian cancer significantly and can provide for up to a sixty percent risk."

Okay, annual checkups for the ovaries as well.

"But it's a fact that the ovaries are not as easy to check as the breasts. And ovarian cancer is much more difficult to treat than breast cancer. The mortality rate is quite high. In fact, they recommend removing both ovaries preventively."

I hadn't thought about that.

"The extent to which they provide hormone treatment to avoid premature menopause is a slightly ticklish question," Gerdes says. "Estrogen is one of the things that stimulates the development of breast cancer. On the other hand, studies show that the removal of the ovaries in women with BRCA1 mutations helps protect against breast cancer."

Still, I would like the test. When it comes right down to it, I have no children, so at least there won't be others who will be touched by the result.

"So, we need to look at your family tree," says Kjergaard, tracing a finger from my mother to her brother, my uncle, and then from him to his daughters, my two cousins. "If you have inherited a mutation from your mother, there is a fifty percent chance that her brother carries the same mutation. And if that is the case, there is a fifty percent chance that your cousins do as well."

"And you have a brother," her boss adds. "He might also carry a BRCA mutation that we find in you and, in him, it might give rise to an increased risk of prostate cancer that he can pass on to his children."

I burble something about contacting my brother if it should become relevant. My cousins, as well.

"Yes, because that is something we can't do," Kjergaard says, getting up. "So I'll order a blood test," and it's as though there is something final embedded in her words. "You can go down and have a blood sample taken. Then, we'll call you in for consultation when the genetic analysis has been done, which could take a couple of months."

She sits down to enter my data into the computer system but then abruptly turns toward me.

"When you hear from us, you won't be able to see from the wording of the letter whether there is something wrong or not – we only talk about that face to face."

I shake hands with both of them, and everyone smiles, unconcerned. But I think to myself how the little "see you later" that finally slips out sounds kind of strange.

THE WARD WHERE blood samples are taken on the ground floor runs like a well-oiled machine. You take a number and sit down beside the other patients, with their greyish skin, hospital gowns, and portable stands for intravenous drips. When your number comes up, you are called into a little booth, where a frighteningly efficient lab worker pokes a hole in your vein and drains the necessary blood. It takes at most two minutes, and not a word is exchanged before you're sent off again.

Outside in the real world, I forget my bicycle and get almost halfway home before I think of it.

Later, I run through my conversation with the two doctors and play out the different scenarios. Suddenly, it's as if the only realistic outcome of the test is that I *have* BRCA mutations and therefore have to take a position on the available options. In my inner ear, I can very clearly hear Kári Stefánsson say "bilateral mastectomy" without the slightest wobble in his voice.

But could I voluntarily go through the whole thing, the

breast operations and the removal of my ovaries, not to mention the consequences of early menopause? To get cut up crosswise to Sunday and come out at the other end ready for grey hair and liver spots? On the other hand, how would it be to live in the constant fear that the next check-up will reveal the insidious disease?

At night, I'm seized by some bizarre ideas about the two BRCA genes and the proteins the genes code for, whose job is to carry out ongoing repairs to the genome. In a series of cartoon-like images, I envision how chromosomes sustain harm. When the back of the long DNA molecules is broken, a system is set in motion to put the strands right again; BRCA1 and BRCA2 take on this heavy burden. But if the genes are mutated, the proteins are handicapped. It is like being assisted at the roadside by a blind car mechanic with one arm. Sometimes it goes well, but sometimes the repairs go wrong, and very rarely something so harmful will happen that it destroys some of the control mechanisms that keep the cell from growing wildly. The result is the first uncontrollable cancer cell, which just keeps growing and growing.

When these ideas start metastasizing, I repeat a saying that, in some way, always seems comforting.

Death must have a cause.

It was a slogan repeated time and again in my family. We were not afraid of dying, because the consensus was that, in the end, it was not death but life that was a vale of tears. As my father used to say with a wry smile: "I've got no problem with death, I just hope there's no reincarnation."

Thus cheered, I remind myself that none of us are free of mutations and genetic weaknesses — the flawless genome does not exist. This is a truth that will become blindingly clear as more advances mount in the world of personal genetics. The more genomes are mapped, the more people there will be who purchase tests and gene profiles, and the more accepted the idea of the imperfect will become.

Of course, you might be able to turn angst and eternal worry upside down. You could ask whether we might use our familiarity with our personal dispositions to illness to reconcile ourselves with the idea that we will one day depart from this life. I think about this quite a lot.

Sometimes the waiting seems infinite and unreasonable. Because I know that sequencing the two BRCA genes takes no longer than a week with today's technology, the system's two-month waiting period seems especially outrageous. Finally, I call Kjergaard to ask whether something might happen soon. She passes the question on and reports back that the answer will come next week. She probably also senses a certain anxiety, because she offers to save time by notifying me about the appointment for our final conversation by e-mail instead of by letter. I thank her.

The e-mail arrives three days later.

"Your results are ready. You may come by my office on Monday at 3 p.m. We recommend that you bring a relative to the consultation."

It's Friday. All weekend I'm plagued by diarrhoea.

"YOU LOOK A little stressed," says the chief surgeon, placing a piece of paper with two signatures in front of me. "But, fortunately, I have good news for you."

Response to mutational analysis for hereditary mamma (ovary) cancer (BRCA), it reads on top in large letters. Below the conclusion is written in small letters: No pathogenic mutation in BRCA1 or BRCA2 has been detected.

I've grown accustomed to the tight feeling in my gut, but now it dissolves and makes room for a sort of bubble. The office seems warm and inviting, and even Kjergaard smiles broadly and persistently. I breathe out demonstratively.

Fantastic! I've avoided the statistics and my family curse, I say foolhardily. No BRCA mutations but, to the contrary, an analysis

from deCODEme that points to a smaller risk than the average –
I'm in seventh heaven.

"No, no. You can't look at it that way," Kjergaard emphasizes at
once. "We can't know what the situation was in your mother and
grandmother. It could easily have been a flaw in genes we know
nothing about yet, and which has been passed on to you."

She stresses that the medical system's recommendations are still
on the table: regular checkups with the breast surgeons. That's fine
– nothing she says takes away my feeling of being twice acquitted.
With the mention of deCODEme, however, I have reminded
Kjergaard of a development she does not care for and which she
now begins to discuss. She meets regularly with people who have
heard about some chromosome flaw or mutation and go onto the
Internet to read up on it and become very frightened.

"They simply do not understand the information," she says
irately. "And what happens to the healthcare system when we have
to advise a stream of people about various tests they have purchased
from some private company and can't interpret themselves? Where
are we headed?"

Offhand, I can't answer her, but I can easily see the reason for
her objections. Right now, the prospects are foggy, but you can still
sense that something has to happen in the healthcare system,
because a new reality is going to flood it.

Personal genetics is still in an embryonic state, and people can
easily criticize the existing tests and gene profiles for providing too
little usable knowledge. But we are also only dealing with the tip of
the iceberg. Research – public research – is aimed at developing
personal medicine, that is, prevention and treatment adapted to the
individual's physiology, which, once again, has to do with the individual's genes. In the long term, there will be a general consumption of information and services based on genetics with respect to
both diagnosis and treatment. But this is a consumption that, for a
great many of the public, will take place directly – at the consumer's
initiative and away from doctors and hospitals. The market is

private, prices are dropping rapidly, and the supply will only become more extensive. If you put this together with what's happening on the Internet, you get a potent mixture.

It's not just that everyone will eventually be informed of any possible infirmity and its treatment on the great, omniscient Web, but there will be services such as Microsoft HealthVault and Google Health, where you will be able to sign up for an account and use it to keep your own running medical record. If you get sick, there will be organizations such as PatientsLikeMe, where thousands of patients with particular chronic diseases have congregated into a close social network. And they don't chat their time away. They monitor their symptoms and the course of their illness in a professional way; they contact researchers, experiment with treatments, and advise each other. In some cases, they even put together research projects completely outside the healthcare system.

Patients aren't what they used to be, and the doctor's role is also being transformed. As opposed to the good old days, when doctors were everyday gods, all-knowing experts who charted the course of your health unchallenged, the physicians of the future will be service technicians. Middlemen in white gowns who have the necessary authorization to provide the services the healthcare system offers but who, in many instances, know no more than the patients do. Perhaps even less. Right now, in the United States there is a violent debate about what should be done to reduce doctors' ignorance about genetics. Studies show that most "ordinary" doctors do not know enough about genetics to be able to advise their patients on the content and significance of various consumer genetics tests, and specially trained genetic counselors are few. On the other hand, this profession is one of the fastest growing in the world. While it should be applauded that people are taking responsibility for their own health, it must also be acknowledged that the development of personal genetics pushes this responsibility further and further into each person's own hands.

"Where are we headed?" The chief physician's question rings in my ears. I decide to look for those who might have an educated guess about it: the people who are driving the development. It appears that this embryonic industry and some of its cutting-edge researchers are holding the world's first conference on consumer genetics. So I've bought a plane ticket to Boston to hear what they have to say.

4

The research revolutionaries

The revolution is not an apple that falls when it is ripe.
You have to make it fall.

CHE GUEVARA

I'VE COME TO the sprawling Hynes Convention Center to attend the first international Consumer Genetics Show. The atmosphere is breathless. I'd blame arctic air conditioning, but it's obvious that everyone here feels they've been granted intimate seats at the birth of something magnificent.

"This is consumer genetics from every conceivable angle," says the conference's organizer, the biotech entrepreneur John Boyce. I see many people I know from my circuits around the genomics revolution. Sorenson Genetics is here to talk about Americans' increasing consumption of genetic paternity tests, which can be purchased by post or at the local pharmacy. DeCODEme, of course, is also in attendance. Its stand, by coincidence, is located next to a couple of competitors that offer to sequence a person's entire genome, not just a measly million SNPs. Talks cover everything from whether consumer genetics is the business model of the future to how to broadcast your genetic information to the world.

But there are snakes in paradise. Inside the frigid lecture hall, a panel is discussing the "forces" trying to limit or fence in an otherwise free market. The authorities in Germany, for instance, do not believe it is defensible or desirable to promote advanced genetic tests directly to innocent citizens; since not everyone is capable of understanding genetic information on his or her own, any sort of genetic test must in the future be prescribed by a doctor.

"The Germans are an interesting people," a drawling voice announces from the speaker's podium. It is Kári Stefánsson, and he is apparently in a sarcastic mood. "They are happy to sell you tobacco and alcohol and fast cars, which they *know* will kill people, but you mustn't go out and get knowledge about your own risk of disease. That I find a bit strange."

"Arrogant shit," hisses the person next to me, so half the row can hear it. A couple of people nod in agreement, while Stefánsson continues his tirade.

"Generally speaking, the critique of direct marketing to consumers is interesting. We already know that patients use the Internet and are often better informed than their general practitioner. People are interested in their health and, for me the question is whether we have a responsibility to give them access to the knowledge about genetics that research is producing?"

Stefánsson decides to use the occasion of the first Consumer Genetics Show to share the news – or assert as fact – that his researchers have discovered some new genetic markers that tell us something about the risk of atrial fibrillation. It appears that variants in the gene ZFHX3, and their location on chromosome 16, significantly increase the risk of atrial fibrillation and, by extension, the risk for the most common type of cerebral hemorrhage. These small hemorrhages may not in themselves be fatal but they can gradually destroy so much brain tissue that it leads to dementia.

"In the half percent of the population that has the greatest risk for cerebral hemorrhage, we can see that these gene variants provide a whopping seventy-five percent risk. And, at the same time, they provide a far better prediction than the traditional cholesterol measurements. This sort of knowledge can be used to get people into preventative treatment. Yet, still, many doctors and geneticists maintain their resistance – why?"

I'm sitting in the fifth row and I'm a little shaken. Not to hear about doctors' resistance to consumer genetics, or the new German law for that matter; both were predictable. But this stuff about ZFHX3 is totally new to me. Stefánsson's team published the discovery after I had visited him in Iceland just a couple of months ago. The researchers were moving quite quickly, and quite tangibly. People had a reason to be breathless.

Since Stefánsson's ZFHX3 markers hit the scientific literature, they have been entered into the gene profiles from deCODEme, and I'll now be able to check my situation. And that's how the process will work from now on. Every time a research group somewhere in the world discovers a new connection between a SNP and a biological characteristic, I, along with every other owner of a gene profile, will be able to log in and pull out my raw data to check my status. One day, we might be alerted to new data for the risk of depression; the next, susceptibility to athlete's foot. When you think about it, it's a fascinating way to be connected directly to the forefront of genetic research.

"Stop!"

I am pulled out of my musing by a young man, two rows in front of me, who stands and asks whether Stefánsson really thinks he and his colleagues are sufficiently skilled at explaining to people that the knowledge about risks changes swiftly and sometimes drastically.

"The people who buy your gene profiles this year may find that they have a low risk for something or other. But it may prove they have a high risk next year, because research has advanced and may have revealed that the interim genetic studies didn't hold water. Your results are very uncertain, but I don't think the average consumer realizes that."

The young man sounds upset, but Stefánsson just brushes him off.

"Those are the rules of the game. It's in the nature of things that nobody can predict future findings, but we have to act at any point on the knowledge we have *here and now*."

He reminds me of the cancer researcher Bert Vogelstein, from Johns Hopkins University, who wrote something similar in *Nature*: "Humans are really good at being able to take a bit of knowledge and use it to great advantage. It's important not to wait until we understand everything, because that's going to be a long time away."

Aren't we, perhaps, too impatient? A population concerned about health won't tolerate waiting until everything is illuminated to the darkest corners.

Nevertheless, the young man has a point. It is striking how many different messages you can get out of the same genome, not because our genes change but because tests and interpretations do. It is only mentioned in passing at the conference, but there are problems with the interpretations of the risks that accompany genetic profiles, because the providers make use of different gene chips in their analyses. Put simply, they do not test the same variants. Some test seven variants for cardiovascular disease, while others only take four into account, which means the same person can get the tag "high risk" from one shop and "low risk" from another. Some use only nine variants to calculate the risk for type 2 diabetes, while others use eighteen and, thus, are more exact.

Then, there are the updates. Discoveries of significant new SNPs are incorporated with those already known and, just like that, the same genome provides a new risk assessment. In 2009, *New Scientist* asked a group of Dutch researchers to look more closely at the risk assessment for type 2 diabetes developed at deCODEme, and quite a bit had changed in just two years. The researchers did not test living people, but instead produced computer simulations of almost six thousand genomes with different combinations of SNPs related to diabetes – a "virtual patient" database. When deCODEme launched their gene profile in the autumn of 2007, eight SNPs in the gene TCFL2 formed the basis for its risk calculation, but just a year later this was upgraded to eleven, and in 2009, to fifteen. With these changes, four in ten computer-simulated people changed risk categories; one in ten changed categories twice.

Yet would the experiences of virtual patients translate to people in the real world?

Yes and no. You could say that changing risk categories doesn't much matter if the disease in question cannot purposefully be prevented. But if you are talking about a condition for which preventative medical treatment can be recommended for one risk category but not for another, it immediately has great significance.

Of course, consumer confidence is something else entirely. Cecile Janssen of Rotterdam's Erasmus University, the researcher in charge of the study, admits she is worried that what there is might evaporate. If you expect that results, and your probable risk, will change over time no matter what you do, you lose the motivation to react to them. Like when the health authorities advise you to eat something or other to prevent cancer, only to withdraw that advice later.

The skeptics say shut it down, it's far too early, consumer genetics isn't ready for prime time. No, no, no, others argue: the field must have room to develop if it is going to be useful, and it must take place in the spotlight of public scrutiny, where everyone can see what's happening. Consumers must be prepared to accept uncertainty as a fundamental condition. We just have to get used to the fact that science does not always provide absolute, final answers; it is an eternally forward-moving process that constantly changes our understanding of the world.

As Robert Cook Deegan, the director of Genome Ethics, Law, and Policy at Duke University, says from the podium: "No matter what the problems are now, we can't get around the fact that the thinking in this area has itself changed fundamentally. There *is* access to personal genetic information, and the information keeps getting cheaper and better. "

He calls the German law "pure idiocy."

"Wanting to protect people against consumer genetics is like wanting to shut down the Internet because there is porn out there that can corrupt unguarded souls. That's what the Germans are doing with their prohibition. They are acting from fear and without any regard for the potential advantages they are missing out on."

Still, generally speaking, an interesting step has been taken. For many years now, a mantra has been chanted about "toxic knowledge" and the "right not to know." The predominant attitude has been that you should ultimately protect "people" from knowledge that might make them uncomfortable, because as non-experts they are incapable of taking a position on it. A kind interpretation would label that as paternalistic guardianship. Or we could straight out call it the superciliousness of the experts. The first studies of how people react to genetic knowledge are only coming out now, and they indicate that ordinary consumers are extremely good at handling such "toxic" information about themselves.

For years, Robert Green has been researching Alzheimer's disease at Boston University. Among many projects, he and his colleagues have tested relatives of Alzheimer's patients for the ApoE4 variant, which increases tenfold the risk of developing the disease. They found, quite surprisingly, that those who prove to have the – perhaps fateful – variant do not have a higher stress level and do not have a greater fear of the future than those who have not been tested and therefore do not know their risk.

Admittedly, for the first six weeks after receiving the results of the test, the carriers of ApoE4 reported feeling more stressed than people without that certain knowledge, but a year and a half later, when the researchers spoke to them again, there was no difference in the psychological well-being of the two groups. On the contrary, some of those who knew they had the ApoE4 variant paid far more attention to planning their time and thinking about their lives than those who might have the same risk of Alzheimer's disease but had decided not to learn their status.

What then of the gene profiles and the range of disease risks associated with them? According to a study done by researchers at the Scripps Translational Science Institute, these tests do not seem to induce heightened anxiety. For six months, the team followed more than two thousand people who bought a SNP-based gene profile from the company Navigenics; there was no indication that the results caused the customers any distress. This prompted the leader of the study, Eric Topol, to tell the *New York Times*: "Up until now there's been lots of speculation and what I'd call fear-mongering about the impact of these tests, but now we have the data."

Another interesting study, headed by Colleen McBride of the National Institutes of Health, investigated smokers and lung cancer. Smokers related to patients with lung cancer were offered a test for a gene variant known to increase the risk of this cancer in smokers. The hypothesis was that people who learn they do not have an increased risk of a given illness, or that they may even have a lower than average risk, will react with indifference. In other words, they will continue smoking if they think they have a genetic "free ride" when it comes to lung cancer and they will happily gobble down mountains of fattening fast food if their genes indicate the same for cardiovascular disease.

For ethicists, this worry sounds plausible, but the research appears to undermine it. In an extension to the NIH study, no difference was detected in the motivation of those tested to stop smoking, whether or not they had the risk-increasing mutation. McBride sums up the two studies: "The findings may help us put a damper on these paternalistic concerns that we have to protect people from this type of information."

Paternalism has its dangers, too. Consider the disturbing story shared by Kári Stefánsson, in one of his panel discussions. In a study of genes associated with breast cancer, deCODE Genetics identified 110 Icelandic women who, based on the information contained in their family histories, it deduced had potentially lethal BRCA2 mutations, which give the carrier a seventy-five percent chance of developing breast cancer. This is a piece of knowledge that can genuinely be called toxic, and the researchers passed it on to the country's health authorities.

So, a bunch of civil servants are sitting with this information that touches on the life or death of real people – some of whom they might even know personally. What do they do? Nothing. No one contacted the women. These BRCA2-positive women were walking around in ignorance, while their genes were bandied about at a conference at which top researchers mingled with representatives from personal care product manufacturers and health insurance companies. The women may be enjoying a happy ignorance, but wouldn't you be seriously unhappy and disappointed with the health authorities if such information were withheld from you? I try to imagine the day they finally receive the news, and shiver in the cold auditorium.

Once, the most pressing issue when conducting research on people was to ensure they were anonymous and remained so. Subjects volunteered information on the condition that they would never themselves benefit from it and that no one would provide them with their individual results. Now, this old record is sounding a little worn and particularly out of tune with the times.

"I believe the most important ethical, legal, and social question to be studied in the field of genetics is the delivery of genetic results to the participants in the research," writes Catherine A. McCarty of the Marshfield Clinic Research Foundation, in the *Genomics Law Report* blog. In the same forum, geneticist Daniel MacArthur of the UK Sanger Institute offers a hypothetical story of a woman who participates in a diabetes study, and the researchers find she has BRCA mutations. It would be crazy not to give her this information, right?

THERE'S SOMETHING ELFLIKE about Linda Avey, daughter of a Lutheran minister from South Dakota, and one of the founders of 23andMe. It is not just her slender figure but her smile, which spreads effortlessly across her entire face, and her affable manner. You quickly feel she could be your close friend, someone to chat with, someone to confide in. Dark suits whiz along the corridors and, at one point, she leans in toward me and declares in a stage whisper, "Genetics is such a male-dominated world, isn't it?"

That's hard to deny. The field seems a little pumped up on testosterone, bloated by male egos and their hi-tech toys. This may be the chief reason that 23andMe has aroused so much attention – the company was started by two women. Avey admits they are shaking things up by their very existence.

She's not shy about challenging the status quo. "We need a more up-to-date way of thinking about ethics. The research community has pursued the idea of 'protecting human guinea pigs' to the point of paternalism run amok," she says. "We also need a whole new kind of research."

That is a grand ambition. I remind her that 23andMe has often been called genetics's Facebook – and usually not admiringly. Consumers don't just buy a gene profile from the California start-up but also a gorgeous and user-friendly website, where they can invite other users of the 23andMe service to join in conversations and compare their ancestry and risk of disease. There is a special group for mothers-to-be, where they can keep one another updated on the progress of their pregnancy and compare symptoms and genetic information. Of course, some discuss whether their children should be gene-tested after birth.

"The 'Facebook' format is quite deliberate," says Avey, giving my arm a gentle pat. She explains that the choice was meant to encourage information-sharing among users and, in this way, create the basis for a greater change in people's thinking about the research. "It's not popular everywhere. There are a few, but very vocal, top researchers who seem to feel threatened by the very idea of democratizing DNA. Recently, I was at a meeting where one of them became so upset that he almost shouted at me, 'You're trivializing genetics!'"

Avey bulges her eyes and shakes her head in a parody of the grumpy old man.

"No, we're not," she continues, "but we are bringing genetics to the people, and we want to make research far more appealing to ordinary people. Because we believe there is a need for research in which the subjects are themselves actively engaged and they get personal knowledge out of the projects."

Whether or not Avey is proselytizing, this seems like an almost inevitable occurrence in today's technological and social environment. First are the ever-cheaper tools for identifying genetic markers — from SNPs to gene sequences to the map of the entire genome. Then there is the fact that, in the age of Web 2.0, people are comfortable sharing even very personal data. Internet users no longer sit passively and get content shoved down their throats, but instead find it, share it, even create it themselves.

"This idea was a part of our vision right from the beginning, and it grew out of my frustrations from many years in the medical industry," Avey sighs. She looks into my eyes. "What is the biggest obstacle to medical progress? The challenge of all challenges?"

For the moment, I can't think of anything to say, so I smile knowingly, as if this were a rhetorical question. Avey continues, apparently without noticing my evasion.

"It is choosing the right patients for each study and then to get enough of them signed up for the studies necessary to develop and test new medicine."

I nod. From working in biotechnology labs, one of which was a US-based start-up in the field of neurodegenerative conditions such as Alzheimer's and Parkinson's diseases, I know that testing and research on people is colossal, resource-demanding work. Clinical trials, starting from tests on animals and working through many human phases to assess safety and efficacy, count for a huge portion of more than one billion dollars it costs to develop a new drug from test tube to market.

Avey continues, a frustrated expression etching her face. "But the whole research culture is on the wrong track. You have all these university research scientists who feel like they *own* the disease they are researching." White-smocked academics, who petulantly insist on *my* atrial fibrillation and *my* sclerosis. *I'm* going to solve the problem.

"This sense of monopoly prevents the exchange of data, which slows everything down," she says. Perhaps that is why, in September 2009, Avey chose to step down from 23andMe to launch the Brainstorm Research Foundation, which is devoted to collecting the phenotypes and health outcomes of people with genetic markers, including ApoE4, the Alzheimer's disease marker that both she and her husband carry. "Having experienced this delay, it gradually became clear to me that the patients have to be at the center and that they should be the driving force in research and directly involved in it."

Lord knows, Avey's message sounds warm and wonderful, but it's not stopping the gentlemen in the hall behind us from chattering on about whether the public can handle taking a bite from the fruit of the genetic tree. Whether ordinary people will understand the uncertainty that follows from the fact that genetic data, in these early days, are far from exhaustive and subject to constant and intense development.

"Uncertainty exists, and you have to be completely honest about that," says Avey quickly and, perhaps, dismissively. "But listen: we see ourselves as creating a sort of ecosystem of patients and users. People who stay with us on the website, who keep up with developments and continue to enter in their data as they gradually get older. Can you see it? These groups – or cohorts – have built-in opportunities for conducting long-term studies that run for years. Studies that you cannot scrape together the money or the research subjects for today."

Yes, I can see it, and I even think it's a good idea. You could follow the treatment people receive and discover unintended side effects of drugs that no one would otherwise have registered. Or you track the ways people live and uncover how they add up to differences in health, lifespan, maybe even life satisfaction. All these things would, of course, be evaluated through a genetic prism, understood through the connections between diseases, lifespan or other traits, and a person's gene variants.

"And," says Avey, "everything is done on a voluntary basis and absolutely free of charge to the public health system."

I begin to wonder about what this would mean for the user, the individual. Would participating in a genetics-based "Facebook ecology" be a way to engage more intimately with decisions about health and life in general? Time will tell.

To start, 23andMe is tackling Parkinson's disease. In a project supported and funded by a number of Parkinson's foundations and patient associations, the company invited ten thousand patients to receive a gene profile almost free. In return, the patients volunteer to provide regular, detailed updates about their medical history. They are asked to relate what medicines they are taking and what effects and side effects they experience. But they also have to reveal how they are living in a broader sense: their diet, their physical exercise regime (or lack of), and a long list of other factors that researchers can compare to genetic markers to gain greater knowledge about the disease.

"Mind you, we're not just talking about information for researchers, and that is the point. This is all knowledge that the patients and their doctors can compare directly and use to optimize treatment for the individual. There is a need for more concrete knowledge – *data*, quite simply – in everyday medical practice."

On the heels of the Parkinson's project, the company also started its so-called Research Revolution program, for which Avey and her cofounder Anne Wojcicki are hoping to recruit the vanguard from the consumers attending the genomics show (or reading newspaper articles about it).

The Research Revolution – that sounds impressive; I have to admit that these two entrepreneurs are very good at turning a phrase. Their revolution – which will pave the way for Avey's ecological system – is being presented as an essential component in the radical democratization of science, a sort of genuine research movement for the people. On the Research Revolution website, there are ten diseases to choose from; as a user, you help decide which of them you think 23andMe should tackle next. The illness that most quickly enrolls a thousand volunteer patients will immediately get its own research project. You vote with a click and, at the same time, put your body to virtual service.

ONE OF THOSE who jumped at the chance to be a part of the revolution is blogger Jen McCabe, whom I meet by chance in cyberspace. I've collapsed into a chair in front of the computer and am surfing around lazily, trying to get sleepy enough to go to bed, but McCabe wakes me up with the intense pitch of her posts. McCabe has not only reported for duty at 23andMe but has decided to open her gene profile for the first time while the world wide web looks over her shoulder. At home in her apartment, she has rigged up a slightly shaky web cam and, with her face right in the lens, describes how "... cool it is to be able to give something back to the community." This is the enthusiasm of youth in full glory.

Her seven-minute-long video has been seen a few thousand times, so she may already be on her way toward recruiting the next wave of research revolutionaries.

It starts somewhat jumpy. "Oh, God, *wow*," exclaims the well-scrubbed woman several times. She seems overwhelmed. "I'm surprisingly nervous, actually," she confesses. She has waited two days since the notification that her results were ready arrived in her

in-box. "I didn't think I would be nervous about sharing this information, but yesterday I made a list of the pros and cons."

She holds a handwritten placard up to the camera. Under "pros"
you can glimpse words such as "transparency" and "patient-driven
research." Under "cons" she has very naturally written "privacy,"
but I cannot understand why "dad" is also there.

"Well, but now I'm doing it, I'm going to look at my results
for the first time so you can see how I react." Very considerate
of her.

"Wow, oh my gosh," she says again. "Look at this. I have a gene
variant that increases the probability that I can't tolerate gluten
because of a deficiency in the enzyme alfa1-trypsin. Wow. This is
absolutely news to me." She sounds a little shaken, too, though the
implications of this finding are manageable enough.

"But I carry no variants for cystic fibrosis, and I don't have the
alcohol flush reaction – yes! But wait, they say I can likely tolerate
lactose, and that's not true, I always drink soy."

She points eagerly at the screen, but the camera is too far away
to focus on anything.

"I'm not resistant to malaria or to HIV infection. But look here:
I'm a sprinter!"

It appears that McCabe has a variant of the ACTN3 gene, which
gives her quick muscle fibers, which makes her more suited to
explosive sports than endurance sports. "So, that's why it was never
any use pushing myself to run cross country in school," she says,
smiling uncertainly.

"This is probably really exciting for those of you watching, but I
must admit I was really nervous about this," she says. "But the good
news is that I don't have variants for what I might be worried about
– sickle cell anemia and hemochromatosis. And I wanted to share all
this with you; it's because I think it's important to advocate in favor
of patient-driven research."

McCabe looks into the camera with wide eyes.

"Search for more information on the Net, because this is

incredible. I have never been so interested in checking data since I got my first credit card records."

Here in the weak glow of my screen, I ask whether I should follow this bouncy young American and leap into the great research revolution. It costs less than a hundred dollars to get a gene profile – a tenth of what I've already coughed up for deCODEme – and I'm already a convert. It's easy, too. I just need to navigate to that gorgeous and user-friendly website and click a few times, and they'll send a test kit.

In a way, this is donating our bodies to science. In this case, it simply happens before we die and we get the option of watching the researchers scrutinize our bits and parts for greater meaning. Like a little dose of immortality.

It's late, and I'm content just to dip my toe into the water this first time. I set up a free 23andMe account with Research Revolution voting rights. The list is something of a mixed bag: migraines, psoriasis, severe food allergies, arthritis, celiac disease, lymphoma/leukemia, multiple sclerosis, ALS, epilepsy, and testicular cancer. These are all diseases that deserve every possible resource, but I end up casting my vote for the only one with which I have a personal relationship: migraine. Not that I myself have had any terrifying attacks, but an uncle and a cousin have seriously suffered. I know it's a nasty condition. And it helps my resolution that migraine is already number one in the race with 216 reported patients. Number two, psoriasis, has only 99.

Strange that it feels good to check it off. It really does feel like taking a small role in a growing movement, or perhaps like taking part in the future. But does it work to ask people to provide data themselves? Will they keep accurate records? Will they understand the scientists' questions? And will they answer these questions honestly – especially when it comes to such tricky topics as diet, drinking habits, or following doctors' instructions? Self-reported data can make researchers grind their teeth. Can 23andMe find something interesting in its non-random volunteers that traditional research groups, with their randomized studies, real subjects, and recognized universities, cannot?

"I'm a believer," writes Daniel MacArthur on his blog, *Genetic Future*. He puts weight on the effective viral marketing by which people who are already part of the project bring in friends and acquaintances. "As patient cohorts get ever larger and 23andMe's analysis becomes more sophisticated, there's every reason to expect that this approach will eventually yield the power required to generate novel associations."

And if Google, which has married into the enterprise, continues its generous support of its finances for another few years, MacArthur sees great possibilities.

"I wouldn't be shocked to see this research model eventually gather larger cohorts than even the largest academic consortiums, particularly for less common diseases with particularly strong grass-roots activists."

THE FIRST RESULTS of the research on the volunteer users are made public not long after MacArthur's comment. In Hawaii, where the American Association of Human Genetics is holding its annual conference, 23andMe's principal scientist, Nick Eriksson, unveils the findings.

As Eriksson talks, people in the hall hold their breath. It comes to light that the research team, along with partners at Stanford and Columbia University, has identified new genetic traces for three human characteristics. What have they found? Have they finally tracked down definitive data on the genes involved in Parkinson's or something that indicates a cause for testicular cancer? Not exactly. They have found two SNPs that are strongly associated with curly hair, a single SNP that is linked to the sneeze reflex that some people experience when they are exposed to strong light, and a SNP that seems to increase the risk of *aspargus anosmi*, the condition that makes you unable to detect the unpleasant smell of the sulfurous compound methanethiol, which is formed in urine after digesting asparagus.

"Much of human genetic variation remains entirely unexplained," states Eriksson's team. To remedy this, the researchers have focused on twenty-two "ordinary traits," as these are called in the presentation, the genetics of which no one else has really shown any interest in. Using questionnaires, Eriksson and his colleagues have determined if volunteers are right- or left-handed, if they have a dominant eye, if they have ever had braces on their teeth, if they have had their wisdom teeth removed, if they get carsick on drives in the country, if they have an optimistic temperament, and if they prefer to exercise in the morning or in the evening. Somewhere in the vicinity of ten thousand volunteers replied to the questions. Then, the researchers only needed to go to the computer and compare their answers with the information they already had about the users' genes – a gene profile covering half a million SNPs. To answer the skeptics at the Hawaii meeting, Eriksson asserted that the method itself seems to work: the researchers identified a series of familiar genetic associations in their data – that is, clear links between SNPs and physical characteristics such as hair color, eye color, and freckles.

Erikkson had proved that you can, apparently, conduct accurate research on self-reported, user-generated data. But *should* you? Isn't it an appalling waste of time, money, and scientific creativity if you spend these efforts on what makes hair curly or what makes someone bothered by the smell of long-digested asparagus? Weren't there more pressing questions that could have been posed to the nearly ten thousand volunteers? Isn't this exactly what critics of consumer genetics disparagingly call "recreational genetics?"

It could be argued that the whole point of genetic democratization – that anyone may someday be able to afford to buy an insight into their genes – is that the patients, that is, the *people themselves* will influence how the information is used, in research and in life. Who says that serious diseases are the only things you're allowed to be interested in, and that disease genes are the only things serious enough to merit attention? Why shouldn't everything be fair game? Ultimately, consumer genetics is about exploring and discovering yourself at a molecular level – and whatever the consumer demands, the market supplies.

"ARE YOU HERE, too?"

George Church, a professor at Harvard University, seems mildly surprised to see me at yet another conference in Boston. A short time ago, at the Consumer Genetics Show, I exchanged some polite words with him, and here we are again with nametags and cups of coffee. This time, the setting is Microsoft's elegant building in Cambridge, and it's much more exclusive. In fact, it costs one thousand dollars – twice as much as the market price for the cheapest SNP profile – to get in, unless you have a free press pass. The focal point is Church's own special version of consumer genetics – the Personal Genome Project.

PGP, as it is called among friends, is not only highly personal, it is also wildly ambitious. Its vision is to enroll one hundred thousand volunteers who will, free of charge, have their total genome sequenced in exchange for putting it and a broad range of their health information on the Internet. An enormous data reservoir that will make it possible to look for links between genes, environmental factors, and human traits. Data that includes the name and a picture of the participant and is freely accessible to anyone who might want to look.

"I'm inspired by the Wikipedia model," says Church, shoving a quarter of a bagel into his mouth. For him, the project is an attempt to create a biological parallel to the computer world's "open source" movement, where software is free and any volunteer can make improvements and increase overall knowledge. The particularly innovative thing about the Personal Genome Project is that pure amateurs will have access to the data, too.

"These are extremely valuable data, and it would be crazy to restrict them to businesses or academics. The thing is that we can't know who the future innovators are. The technology is inexpensive enough that everyone can be a part, and the next Bill Gates or Steve Jobs could easily prove to be a fifteen-year-old kid who gets some ingenious idea by rummaging around our database at home in her room."

Church himself was one of those people who started his career by building computers in the playroom. Now, many years and a number of inventions later, he has been named by *Newsweek* magazine as one of the world's "10 hottest nerds." The American journalist Carl Zimmer calls him "arguably the smartest, most influential biologist you never heard of." Right now, chomping away, with his wavy hair and a big beard, Church reminds me more of a jolly lumberjack. I remark ingratiatingly that his project makes 23andMe and their Research Revolution look like a May Day rally in the rain.

"Yes, well, we are different from any other project," he says. It is not just the large number of participants that sets his vision apart, but how thoroughly they are measured and analyzed. The Personal Genome Project incorporates health data from volunteers – their health issues, what medications they take, their diet – and then conducts follow-up studies. For example, the research team sends volunteers for brain scans to better understand the structure and functionality of that organ. It also looks closely at each person's immune system, and how the person reacts to various infections. Finally, they take a skin biopsy, which is transformed into immortal stem cells and preserved in a biobank, from where anyone with a strong research proposal can order a batch. As Church points out, "You can get a little piece of Steven Pinker." To investigate how the mind works, I wonder?

You can get a piece of George Church himself, for that matter.

Both he and the celebrity psychologist, Pinker, are among the first ten named participants – the pioneers – who also include the Internet guru Esther Dyson. On the project's website, you can learn, for example, that Church is adopted and dyslexic, while Pinker is of Polish-Jewish extraction on both sides, suffers from esophageal spasms, and takes cholesterol-lowering drugs and folic acid as a dietary supplement. The fifty-eight-year-old Dyson lets it be known that she "... has pretty much never been sick and never missed work." The only thing she permits herself is regular use of sleeping pills and a daily intake of the hormone estradiol for her menopause symptoms.

At first glance, the project seems taboo-breaking. A sort of genetic exhibitionism. How can they do it? And how do they think they can recruit hundreds of thousands of others to perform the same information striptease?

"At the moment, there are fifteen thousand people in line," says Church, a revelation which shakes me a bit. "First, they have to take an entrance exam that shows that they understand enough about genetics to know what they're getting into."

We are talking about ordinary Americans who, according to Church, fall into three categories: people who believe they are unusually healthy and want to help medical research by making themselves available; people who believe they are terribly sick and, therefore, have nothing to lose but everything to gain; and, finally, people who are genealogy fanatics. They have been tested by the Genographic Project and by 23andMe, and they just can't get enough.

"These guys are experts; they know more about genetics than I do," he says as he crushes an empty paper cup. I ask – half in jest – whether he has room for one more, but am rejected because of my nationality. The project only has official authorization to use American citizens. But there are major international plans to collaborate with centers in other countries, which will then start their own genome projects, following the same pattern. In fact,

South Korea is already in the offing, as the second personal genomics project to open its doors.

—◦◦◦—

"TODAY, THERE ARE thirteen complete, named, and publicly accessible genomes," says Church, pointing toward the auditorium, where two hundred participants are gathering. "But this is the first and last conference at which we can gather everyone with a complete genome sequence in one room."

Some of the thirteen pioneers have decided to stay at home. Steven Pinker, for instance, is not here, but James Watson, the pioneer of all pioneers, has flown in from Cold Spring Harbor. The old man seems to be in fine form, wearing a well-fitting tweed jacket, and with a young assistant to look after him. As the first on the stage, he directly addresses Church, stating that he needs to get on with things and sequence some more genomes at the Harvard lab.

"No more talk, just get it done!" he grunts.

Giggles and hushed whispers spread among the spectators, and the day's interviewer, the radio host Robert Krulwich, hurries to the next point on the agenda. The experienced science journalist does not hide that he is skeptical about all this publicly accessible genetic information, but he starts gently.

He turns to Henry Louis "Skip" Gates Jr., the Harvard professor of African American Studies. He and his ninety-seven-year-old father have both been sequenced: "the first African Americans and the first father-son pair," as Gates notes. He adds that if anyone in the room has not seen the television series that came out of the project, we should buy his DVD. He then continues in a less business-minded voice. "First and foremost, I wanted to immortalize my father," he says.

From my seat at the back of the auditorium, I can't help but think that this jovial man is probably best known for his scrape with a white policeman, who thought the professor was breaking into his own home in upmarket Cambridge. "Racist" was shouted by one side, "arrogant academic" by the other. The matter had to be resolved by President Obama, who invited the parties to a media-transmitted conciliatory beer in the White House garden.

"It was deeply moving," Gates says, referring to a scene in the television series in which father and son look at their respective genomes for the first time.

"The most fantastic thing was that we could separate my father's sequence from my own and thereby get that part of my DNA that came from my mother. It was like having my mother brought back. Dad cried, and I was on the verge of it myself. It was powerful."

"Why is that?" Krulwich wanted to know. "A DNA sequence is just a printout with graphs in different colors."

"What is a photograph?" Gates asks calmly. "It's 'just' a graphic representation of a real person on paper. The genome is a metaphor for a person in the same way, and just as people once had to learn to look at photographs, we now have to learn to see the person in these graphs and bases. It's all a part of our identity."

Then, the dark-skinned Gates tells us that, by the way, he's really white. His mitochondrial DNA and his Y chromosome point directly to origins in Europe.

"For thirty-five percent of black American men, their Y chromosome comes from some slave owner, and that's a shock for many people. But as far as I'm concerned, genetics has to do with breaking down the myths of racial purity. It's good for us to realize that we are all a human version of bouillabaisse – a wonderful mixture."

Unfortunately, after this feel-good beginning to the day, Gates has to leave for a budget meeting with his dean. Now we hear from a couple of businessmen: Jay Flatley, the CEO of Illumina, and Greg Lucier, who runs Life Technologies. Both companies dabble in genome sequencing, and both directors exude an air of purposeful no-nonsense. Flatley and Lucier have had their genomes sequenced and made public, and Krulwich wants to know whether they are nervous that shareholders and board members will dig into the data and find things they don't care for.

"I'm perfectly healthy," Flatley says plainly.

"Your families, then," continues Krulwich. "How do they feel about it?"

"We've talked a lot about it at home," says Lucier, who is in the process of having his wife, his children, and his parents sequenced. "Everyone is very enthusiastic and, with the business I'm in, we feel it's perfectly natural. We want to be on the cutting edge and be examples to follow." He shoots a gleaming smile at the audience.

"Yes, but the *children* ..." begins Krulwich, who now has sweat on his brow.

"Okay," Flatley says. "It could be an argument for *not* putting your genome on the Internet that your children will automatically have half their DNA made public without having anything to say about it. Of course, that is a question you must consider carefully."

"I just have a hard time seeing what you get out of making it public," says Krulwich.

"Think about the Internet."

At this point, the diminutive, fragile-looking Esther Dyson joins the two compact CEOs on stage. She explains patiently that so-called reasonable people could not see the use of a lot of apparently foolish things that we today understand as essential to the development of the Internet. The Internet we can no longer live without. "In the same way, a lot of strange interests in genetics will be part of driving a development that will ultimately benefit a lot of areas, not least the medical field," she says. "You can't sit here today and see what will be important in the future. Publicity and attention are good, because people need to be learning about this. In ten years, there really will be some applicable knowledge, and it's no use waiting until that time."

Dyson also doesn't understand why it's assumed that people won't be able to grasp all the statistics involved. She respects people's intellect, she says. "If they can understand baseball statistics, they can grasp fundamental genetics. And it is easier to understand, when you are looking at your own data rather than some random textbook."

Flatley interrupts her. "We need to get to the point where the advantages of having your genome on the Net are so great that it would be crazy not to have it there."

By chance, his company has a vision for developing a whole series of products directed at genetic data and designed to meet the public's idiosyncratic interests over the long term – a bit like the many apps you can buy for your iPhone. Indeed, the first will be a nifty little program that makes it possible to explore your DNA *on* your mobile phone. You will be able to search your genome when you're a bit bored or when you're at the doctor's and she is about to prescribe some medication without knowing your tolerance for it.

"Or what about when you're sitting at a bar watching women. You can exchange essential biological data before you go any further," suggests Krulwich, sarcastically. Then he becomes serious. "It all sounds so easy, but isn't it terribly risky to have such deeply personal information out in public?"

"It's strange," whispers the person next to me who, according to his nametag, is Kirk Maxey. "People think of genetic information as being particularly risky. Nobody thinks about the fact that information, for example, about our credit rating is out there in cyberspace. Nobody has asked permission to collect it, and they share it and sell it to the highest bidder without my knowledge or consent. Information about who I owe money to, the extent to which I pay my electricity bill on time, and that sort of thing. Might that not be far more explosive personal information than any gene sequences?"

His question hangs there as the panel keeps talking. But when they break for coffee I ask Maxey for his personal history, which proves to be fascinating. Today, he is the director of Cayman Chemical, and a married man with two children. In his youth, as a poor medical student, however, he was a sperm donor. Not motivated by the money, he emphasizes, but by "a desire to help the childless." That he certainly did. The fertility clinic at the University of Michigan told him that most of the thousand donations Maxey made over fourteen years would be used for research into artificial insemination; later, he learned that he is the presumed biological father of four hundred children, many of whom live in the same area as he does. It came out in 2006, when he was contacted by two of the children, who had taken tortuous paths to find him. Maxey would go on to advocate reforming the entire sperm donor industry.

"Sperm donors should be gene-tested," he says a little too loudly. "Since we have the technology, it is irresponsible not to make sure that they are not spreading all sorts of heritable defects."

Maxey has taken the bold step of becoming one of George Church's ten pioneers, so his genome and medical records can be studied by any interested descendants. He confides, "We have to get used to the fact that, in a time with the Internet and genomic databases, donors *cannot* be anonymous. It is simply unreasonable to deny someone access to half their genetic heritage."

UP ON STAGE, Krulwich is taking on yet another victim. This time it is John West, an American tycoon who, at a price tag of two hundred thousand dollars, has had himself, his wife, and their two teenage children sequenced by the company Knome. The West family was splashed across the front pages as the first family to have their genomes sequenced for non-medical reasons. Krulwich returns to his argument that it's awful to give children this sort of information. "Isn't it a terrible burden to put on them? That they are going to go around pondering their disease risk so early in life?"

West brushes him off. "In a few years, it will seem unethical if you *don't* have your children sequenced. And it will be some pretty poor parents who don't shell out that bit of money for their children, for something that will help them discover things about their own health."

Down in the cheap seats, among the freeloaders from the press, there is a murmur about paying fifty thousand dollars per genome. Even though the price is expected to drop in the years to come, one listener asks whether there are plans for facilitating access to this important information for people other than the rich and their children.

"My hope is that the government will eventually sponsor genome sequencing for all newborns," says Jay Flatley, who ventures a prediction that this will be realistic by 2020. "It is on this basis that we are even implementing something like the Personal Genome Project and having debates like this one."

Esther Dyson does not quite believe such government support will be forthcoming in the United States, which has not managed to pass legislation giving all citizens public health insurance, despite numerous attempts over the decades. On the other hand, Dyson imagines that 23andMe, on whose board she sits, will emerge as the future supplier of DNA sequences for everyone.

"We would like to offer whole genomes with the same type of customer service we have today for SNP profiles. Pretty soon, we won't even do SNPs."

DURING THE BREAK, I'm standing in a corner feeling embarrassed by my passé SNP profile from deCODEme. It almost feels like I'm carrying around a chunky first-generation brick of a Nokia, while everyone else is watching videos on their iPhone 4s.

"But at the present time you can't get much more information from the total genome than you can from the million SNPs that may be on a gene chip," Earl Collier consoles me. Collier, who goes by the name "Duke" and is a director of deCODE Genetics, is fundamentally correct. As he reminds me, they don't even know the biological function of ninety percent of our more than twenty thousand genes. Even though much is being done to develop the facility to produce sequences faster and cheaper, the research providing knowledge and interpretation is limping far, far behind.

"To this point, as far as the vast majority of people are concerned, most of the associations we know of between genes and illness have been found through studies of SNPs. It will be a long time before we get much further with whole genomes."

Still, I'd like to be part of the in-crowd and have a proper sequence, now that so many luminaries have one. That's part of the allure of genetic data. But I'll probably have to wait until the price gets down to one thousand dollars, which is expected within the next three to five years.

"You also have a chance to win a free sequencing today," remarks a young man who is trying to squeeze a cappuccino out of a reluctant machine. He proves to be Jason Bobe, the community manager for the Personal Genome Project and the person who makes the outfit work, in practice. He is also the organizer of the conference. And to give the gathering a little extra resonance, he has announced a competition for the most original idea for how to apply personal genetics within the next five years. The winner can have his or her genome sequenced free of charge. To be considered, your idea must have the potential "to improve one or more aspects of human life, such as health, well-being, reproduction, ability, safety, or entertainment." Oompf.

I have no immediately obvious, original ideas for any of these categories.

"Come up with something," encourages Bobe. "We already have a couple dozen proposals, and the four best will be selected and voted on by the participants via SMS. Just like *Idol*."

The notion of an amateur's paradise takes me far away from the pleasant greenhouse atmosphere at Microsoft's offices. Here, there is a well-informed and well-to-do audience rubbing elbows and speaking easily about "microbiomics," and discussing the finer nuances of "next generation sequencing." How would this discussion sound to an outsider? What does all this talk mean to, say, the girl serving us flaky, fresh mini croissants? Would she pick up her phone and send a text in support of a personal genetics application? Would she care?

Civilization is in "a race between education and catastrophe," wrote the British author H.G. Wells. That race is on, now more than ever. Not only are some powerful new technologies developing at an almost appalling speed, they are also cheap enough and accessible enough for a broad cross-section of the population to use them. Mindlessly and completely automatically, these technologies uncover information previously hidden from humanity. And no one – not even the élite researchers and industrial magnates congregating at Church's conference – has a full view of how these technologies and this information will be used, or misused, over the next few years. Rather, they are eager to place both in our hands to see what new breakthroughs they will bring.

At the same time, *education* about genetic technologies and information is stumbling a long way behind. Neither the average retiree nor the average university student is certain to know what genes are, where they are found, or what they do.

"Public education – especially, for the young – is one of the primary missions for the Personal Genome Project," says a crisp voice behind me. It belongs to a boy in a pinstriped suit. A boy who looks to be about sixteen but who proves to be a lawyer specializing in bio-legislation. "Dan Vorhaus. Here's my card."

I take it, and he continues undisturbed. "With its total openness, the project can provoke the political class and the education system to take the challenge of gene research seriously. And I actually think that the researchers and the industry working in the field have a responsibility to inform people about what genetics can and cannot do."

For Vorhaus, who nabbed a degree in bioethics before he went to law school, it is all about clearing the mystique surrounding genes.

"The general sense is that there is something exceptional about genetic information. That it is in some way qualitatively different from all other types of personal information and, therefore, should be treated differently."

Doesn't he think it should be? After all, we are talking about sensitive information regarding disease predispositions – information that can come back to bite you. Fear of misuse of genetic information has spurred American politicians to enact the Genetic Information Nondiscrimination Act, or GINA. The law took effect in 2009 and is specifically designed to protect citizens from having their genetics used against them by health insurance companies and employers. That is, no one may inquire into or use information about people's genetics.

"With this law, they're just sticking their heads in the sand," says Vorhaus, clearly irritated. Then he clears his throat and straightens his tie.

"With respect to GINA, I don't think it's the answer to how you regulate the use of genetic information. It just establishes a general prohibition against use. We need regulation that makes it possible to use the information where it is practical."

I note, for the record, that the first court case about genetic discrimination began in the state of Connecticut. Pamela Fink was imprudent enough to tell her employer, MXenergy, that she had learned she had BRCA mutations and, therefore, was going to have her breasts removed as a preventive measure. Suddenly, after many years of promotions, praise from management, and fat bonuses every Christmas, Fink was demoted and, soon after, fired. Six weeks after her operation, she was escorted from the building, carrying only a cardboard box filled with her possessions.

"We've only heard about the case from Fink and her lawyer, who went to the media," stresses Vorhaus. "And, in fact, I don't think her employer had the right to fire her under pre-existing legislation."

His real point is different. Because you could turn it 180 degrees and ask whether it might be in the employee's interest to have certain genes tested before you take a job.

"For example, there may be components of the job from which you have a particular risk of getting sick," says Vorhaus. This reminds me that a group of researchers at Gentofte Hospital in Denmark is applying for a patent on a gene test that reveals a particularly high risk for eczema. The researchers suggest that all children with atopic dermatitis undergo the test, so they know whether they should avoid careers involving a lot of water and frequent hand-washing; jobs that would probably give them enormous problems with skin allergies.

"That is an excellent example," Vorhaus smiles. "You could also point to the fact that it would be bad for pilots or operators of heavy machinery to have a strong genetic predisposition for epilepsy. GINA does not take these kinds of cases into consideration. What we need is collaboration between government, employers, and employees, that promotes the sharing of genetic information and, at the same time, ensures that the information will not be misused."

As a genuine genome idealist, Vorhaus is going to share his sequence with the world. But what would he think if a prestigious law firm turned him down for a job because of a high risk for Alzheimer's disease, for example? A disease that might – just might – hit him far into his future.

"Hmm. Yes. Well, if it were a crucial part of their business… " His voice fades, and he starts again. "As a matter of pure principle, you can have the opinion that no one should be able to be punished for their inborn genetics, but maybe this is something the government should take care of instead of putting the burden on private employers."

The young man does not go into what he thinks the government should do. But he may be right that it will take some real-life cases and instances that can be debated before politicians will address the problem in a more informed and nuanced piece of legislation. As he says, "We lack experience of how the information can be used both positively and negatively."

I can easily imagine a lot of negatives in the strange area called "abandoned DNA." We all leave DNA on coffee cups, wine glasses, cigarette butts, and toothbrushes, and it is not illegal for others to collect it, have it sequenced and tested for selected SNPs, and make the results public. In the US election of 2008, rumors abounded that Obama's staff gathered up anything left behind, so no one could steal the candidate's DNA and, perhaps, find something compromising in it. At the time, the bioethicist George Annas and the geneticist Robert Green warned against a future of "genetic McCarthyism"; a scenario in which candidates smear each other with unfortunate DNA information and in which, in practice, it is necessary to publish information about selected gene variants even to be considered for political office.

"Actually, I'm surprised that there hasn't yet been a fanatic or a tabloid that has carried out such a stunt," says Vorhaus, looking genuinely at a loss. "But it will happen at some point, and it will probably involve some celebrity."

I think out loud that somebody could follow Brad Pitt and his flock of children and discover that one or more of them weren't his. You could imagine this plaguing some royals, here and there.

"That sort of thing will definitely provoke legislation. Honestly, it is hard to argue that you alone are responsible for the DNA you inevitably leave behind and that it is everybody's right just to take it."

Beneath all these juicy examples, there is an underlying and fundamental question about the right to "genetic privacy." Whereas over time we have had debates about all sorts of information, we still have not thought through and taken a position on what it means to have genetic information floating around in a public space. What sort of right, if any, should we have as individuals – the right to keep genetic information secret, or the right for this information not to be misused?

"*That* is an interesting debate," he replies with a glint in his eye. "My guess is that it will come up when there are many more published genomes and the topic is more urgent."

It is equally interesting what our relationship with genetic information will be on the other side of such technological development and social debate. All in all, the path is being prepared for some tremendous upheavals. We will help cobble research projects together ourselves and we will be able to upload our innermost biology to the cyberspace commons. It may be the death of the old concept of DNA as something exceptional. People may even come to talk about a social-scientific transformation, a leap to a new paradigm, in which we no longer consider our genetic information, or for that matter cells taken from our body, as *us*. Rather, we'll be talking about tools that can be used by us and made available for others.

This sort of thinking stands in stark contrast to the current view that is expressed in all the stories about DNA piracy. In fact, as Vorhaus and I are standing here, debating visions of the future, a relevant case is being hotly discussed in the media. It involves a group of Native Americans in Arizona, the Havasupai tribe, which has received handsome compensation for the use researchers made of tribal members' DNA, without their clear consent.

In 1990, a number of Havasupai voluntarily donated their DNA to researchers at Arizona State University, who were looking for genetic causes for the tribe's vastly increased risk of diabetes. The scientists found no diabetes genes. However, the donated material was later used for other projects. More than a dozen scientific articles, on topics ranging from the genetic causes of schizophrenia to the descent of Native American populations, were published. It was the latter studies that, many years later, the Havasupai especially resented. Genetic analyses show that the tribe, like many other Native Americans, originally migrated from Central Asia. But this was an affront to the Havasupai's creation myths, which state they originate from the area, at the bottom of the Grand Canyon, where they still live today. The genetic data didn't just rock their spiritual identity; the Havasupai also depend on revenue from tourists who visit the canyon and buy souvenirs tied to the tribe's origins.

In newspapers and blogs, the sympathy is for the Havasupai. And it's spelled out in capital letters that researchers may not touch a

DNA sequence without getting the donor's informed consent for every thread of research they pursue with it. But apart from the fact that this is not practical from the point of view of science, in which a study may ask one question yet find an answer to another, is this a reasonable starting point? Is anyone's DNA, in this sense, his or her own? Couldn't you argue with just as much right that the information in the double helix belongs to the larger common human heritage? And that no individual has a right to stop it from helping others?

"You should think of your genome like it was a cell phone," George Church says suddenly. He has been casting longing looks toward the buffet but decides to add his two cents to the conversation. "Imagine that everyone bought a cell phone and just walked around looking at the shiny thing but never gave their number to anybody, because they were afraid of getting unpleasant calls," he continues. "It doesn't work, right? Telephones, fax machines, e-mail, and that sort of thing are only useful if you share the information, and the same holds true for genetic information. Ideally, everyone should be a part of one genome project or another."

Before sheer enthusiasm gets the upper hand, I recall that the interest is still quite limited out there among "everyone." The publicity juggernaut 23andMe has thirty thousand customers, George Church's Personal Genome Project has fifteen thousand people in line, and deCODEme boasts fewer than ten thousand clients. In other words, we are far from anything that resembles the market for mobile phones or e-mail.

"I think we will see a turning point when it suddenly becomes popular and catches on," says Church, who seems to be considering when this point might come.

"Everything that is *not* about disease can make the difference. If there is anything people are seriously interested in, it is behavior, personality, and brain function. These are things that the Personal Genome Project will make it possible to delve into. The participants will soon undergo tests of cognition and social function," Church relates, reaching for a mini croissant after all.

I am inclined to believe that he has a point. Until now, personal genetics has largely been sold and served up as something that concerns health. Of course, health is important, but when it comes down to it, our diseases and infirmities are not really the most interesting thing there is about being human. All the physical stuff is rather a base. It needs to be in order, but you don't really get turned on by that.

Where it really gets exciting is at the intersection between the shell of a physical being and the *person* we recognize as human. How do we get from genes to what we call, for lack of a more scientific term, the soul?

5

Down in the brain

Whenever you can, count.

Francis Galton

WHAT IS THE most interesting subject in the world? For all of us, it's the same thing – ourselves. *Me*. And one of the most tantalizing questions is: *How did I become who I am?* Its answer is naturally linked to genetics, to the biological heritage that shapes our psyche, mindset, and, ultimately, our life. Is life's trajectory plotted out on a predetermined course, or do we have the capacity to direct it in accordance with our iron will?

"My temperament, my view of life, my personality – genes can't possibly be involved in all that. It's just not true!" Thus spake a female artist I met in Los Angeles. Her eyes widened behind a pair of trendy, heavy-framed glasses. She simply couldn't reconcile herself to the idea that she herself was not the absolute sovereign of who she was; she felt *in her soul* that she was completely free to choose. That one's ethereal psyche was not shaped by wet, sloppy biology but by education, experience, environment – not nature, in other words, but nurture. "It's about who I *am* – it's not something physical," she said.

Yet the thing is we *are* physical. We do not experience the world as it is but as it appears after being filtered through hundreds of billions of cells in our brains. The way these cells communicate with each other and thus react to external stimuli is, in part, determined by genetic specifications. Admittedly, the brain is so complex that its structure is not merely determined by our genes; we can see this in studies of identical twins, who are usually genetic clones but do not have completely identical brains. Regardless, genes help define the development and function of the brain, cell by cell, throughout your life. Busy receptors pass on a steady stream of nerve signals; growth factors govern the ongoing reconstruction of the brain's architecture; enzymes take care of metabolism. The availability and efficiency of all the brain's leading actors are specified by genetic information.

"That genes strongly influence how we act is beyond question," an article in *Science* states, before rolling out a history of the genes that have been closely studied for behaviors ranging from aggression and depression to infidelity and lousy love lives. To judge from the coverage in the major scientific journals, the field of behavioral genetics is the place to be.

This has not always been the case. Exploring the genetic basis for how people think, feel, and behave was unpopular for quite a long time. For decades, the discipline was connected with a queasy-making stretch of science's history. The stench of eugenics and genocide clung to behavioral genetics like cigarette smoke to a wool sweater.

It all started with Darwin's cousin. The British anthropologist and all-around intellectual Sir Francis Galton read Charles Darwin's works on the evolution of species with enthusiasm and immediately called his kin's attention to his belief that human mental faculties must be subject to the same laws as their physical characteristics. Intelligence and character must be heritable. To explore this intuition, Galton investigated the traits of many of the relatives and descendants of Victorian Britain's most prominent and highly gifted men. By tracing family trees, he observed that there were many more "eminent relatives" in the families of geniuses than in the general population, but that the number of highly gifted people fell the further away in relationship you moved from the genius himself. Heritability jumped out at him, an indisputable fact.

Galton described his observations in his book *Hereditary Genius,* published in 1869, a mere decade after Darwin's foundational *On the Origin of Species*. In the following years, Galton pondered how his insight could be put to use. In 1883, he found his solution: the concept of eugenics. In his *Inquiries into Human Faculty and Its Development*, he proposed that British society get its act together and find some incentives to encourage the especially gifted to have more children and, thus, spread their desirable qualities among the population.

An intellectual heir to Galton was the American biologist Charles Davenport, who in 1910 founded the Eugenics Record Office at the nascent Cold Spring Harbor Laboratory. The early twentieth century saw research on heredity begin to blossom, through experiments with fruit flies and other similarly amenable animal subjects. Inspired by his experiences in the lab, Davenport turned Galton's idea on its head, so to speak: instead of encouraging the breeding of more good qualities into the population he advocated preventing the spread of bad ones. Davenport was worried about society's general welfare, and he especially wanted to cure the ills that came in the wake of caring for the mentally retarded, the psychotic, and the addicted. Such individuals must harbor "bad" traits and if they could be prevented from passing them on, they would disappear from the population.

Davenport's program was immediately and eagerly implemented throughout the civilized world. In the United States, "inferior" ethnic groups were precluded from immigration for eugenic reasons, and in many European countries, so-called "cretins" were forcibly sterilized. The movement reached its grotesque crescendo with the Nazi extermination of Jews, Gypsies, homosexual men, the mentally ill, the mentally handicapped, and other "asocial elements." It was this catastrophe that, after World War II, relegated genetic tinkering to the graveyard of ideologies.

But out in the trenches of research, a more vibrant and enlightened understanding of the interplay between nature and nurture gradually arose and, after the discovery of the structure of DNA, studies based on a scientific approach and method flourished. Inspired by Galton's original work, people began to investigate the question of the heritability of mental traits and psychiatric illnesses by studying twins. Since then, twin studies have become a fundamental tool of *quantitative genetics*, a branch of genetics focused on establishing whether a trait has a genetic component and, if it does, how great that component is.

Heredity, or rather the *degree of heritability*, can be difficult to pin down. It is not a number that applies to an individual but to an average in the population. If you can ascertain, for example, that height is ninety percent determined by heredity, this does not mean that ninety percent of my five foot eight-inch frame is determined by my genes and the rest by my diet and general state of health. On the other hand, it does mean that ninety percent of the variation in height you see in the entire population is due to the genetic variation found between individuals in the population.

In mathematical terms, the heritability of a trait is the part of the variance in the trait under investigation that can be ascribed to genes. Twin studies try to estimate this component in various ways. One is to compare a given trait in identical twins who have been separated since birth and so have identical genes but carry the baggage of completely different environmental influences. If the trait varies, it cannot be explained by genes alone. Using statistical methods, scientists can deduce how much influence is due to environment and genetics respectively. Another approach is to compare a trait in identical twins, on one hand, and fraternal twins (who only share half their genes), on the other. Again, the difference between the two groups can be used to calculate a measure for the heritability of a trait.

Twin studies are not without their critics, particularly because the precise methods and mathematics behind them are always under development and up for discussion. Can you allow yourself, for example, to assume that the environment of twins is identical just because they are born at the same time and grow up in the same place? Depending on which mathematical trick you use to correct for that sort of question, you have to face the fact that studies of the same trait may provide different numbers for heritability. Still, this does not alter the fact that the method is an excellent tool for assessing heritability.

If you look at schizophrenia, for instance, you can determine how frequently twins, whether identical or fraternal, are struck by the illness. If there is a genetic component to schizophrenia, you would assume that more often both identical twins would be affected than would fraternal twins. And indeed, the statistics reveal quite plainly that if you have an identical twin who is schizophrenic, your risk of getting the disease is fifty times higher than the general population's. If you have a fraternal twin or another sibling with schizophrenia, your risk is five times the general population's. When these numbers are translated into the equations of quantitative genetics, scientists are able to conclude that schizophrenia is up to eighty percent determined by genes.

The idea that diseases have something to do with "faulty" genes is quite digestible for most of us, but many people have difficulty in accepting that their own, quite normal psychological traits also draw directly upon their DNA. Nevertheless, this seems to be true. Over the decades, again and again, twin studies have indicated that there is some degree of heritability in pretty much every psychological or mental trait and in every possible type of behavior. Even in cases you might not at first believe have anything to do with DNA. The American psychologist, Eric Turkheimer, of the University of Virginia, has therefore proposed what he calls the first law of behavioral genetics: that "all human behavioral traits are heritable."

Take intelligence, one of the most intensely studied phenomena. We are infinitely fascinated with this trait: how we measure it; how we compare to one another in intelligence; whether we excel in one sort of intelligence rather than another. People have fought about and scrutinized the very meaning of "intelligence" for centuries. As it turns out, intelligence, as codified in the standardized test for IQ, appears to be one of the most heritable of normal traits. It looks as though eighty percent of the variant in IQ among adults may be ascribed to genetics. Another mental faculty, memory, is also partially inheritable, but here the rate slides down to around twenty percent.

WELL, YOU MIGHT be thinking, in these instances we're talking about mental ability – that is, a quality that can, to a degree, be traced back to the mechanics of the brain and how well it functions: it is almost too obvious that mental ability must be partly physical. But even particularly complex behaviors, and other "soft" characteristics we often chalk up to psychology or sociology, also display a surprising degree of heritability.

Consider, for example, "compulsive hoarding." The need to fill your house with Christmas elves, vintage comic books, or beer labels is, apparently, distinctly genetic – at least, that is, in women. In 2009, a major study involving over four thousand female twins came to the conclusion that the heritability was nearly fifty percent. Having a hard time understanding the collector's itch? Well, the degree of empathy you are capable of has been shown, in scores of twin studies, to have a heritability of between thirty and fifty percent. And your level of religiosity is not simply foisted on you through parental preaching and regular churchgoing. In a major comparative study of a number of populations published in 2005, researchers at the University of Minnesota concluded that your tendency to be religious or spiritual is over forty percent heritable.

In the political sphere, a small group of social scientists, many based in the United States, has turned to genetics to expand their explanatory models, since humans are biological beings, after all. Among the pioneers are "the two Johns," John Alford of Rice University in Texas and John Hibbing of the University of Nebraska-Lincoln, who in 2005 found evidence of heritability in our fundamental political attitudes. Whether you are conservative or liberal is not just a question of parental influence and environment but also genetics. The researchers asked American and Australian twins a battery of questions about the extent to which they were for or against hot-button political issues such as gay rights, the death penalty, and school prayer. They found identical twins had a far higher rate of replying identically than did fraternal twins – so much so that it appears holding liberal or conservative values is forty percent heritable.

Alford and Hibbings' study was controversial, a call to arms for a new area of research carrying the unattractive name of *genopolitics*. For the past few years, the field has been led by the young and ambitious James Fowler of the University of California, San Diego, who advocates an amalgamation of sociology and genetics into a "new science of human nature." As a result, he has been bestowed with superlatives such as "most original thinker of the year." In 2008, two studies of voting behavior, made on the basis of the voter registration records of eight hundred sets of American twins, found that heredity played up to a sixty percent role in whether people went to the voting booth or stayed on the sofa. A year later, Fowler showed that genetics also plays a role in whether you are a loyal adherent of a political party, independent of which party.

Some observers balk at such findings, wondering how genes can possibly have any significant influence on a behavior that is a modern invention and not the least bit "natural." Nonsense, Fowler replies. If you peel away all the trappings of political slogans, polling booths, and the like, political behavior is about our attitudes toward cooperation and social exchange – human inclinations that were very relevant to our survival in the Stone Age. Fowler imagines the genes that play a role in politics are the genes that have long regulated our social behavior and taste for working in groups.

Fowler admits that tying a gene to a prehistoric behavior involves a lot of assumptions. For that reason, he is not satisfied with the sorts of details that can be gleaned from twin studies. In a way, twin studies correspond to flying around at thirty thousand feet observing the landscape. You can see something is growing in the fields below and register that it is green, but there is no way to tell whether it is wheat or barley. Similarly, the fact that there is something hereditary at play does not tell you *what* is hereditary, that is, what biological mechanisms are involved.

To get into these genetic weeds, scientists resort to *molecular studies,* which search for the specific genes that increase a risk, whether of schizophrenia or conservatism, and identify how the genes act – what enzymes they produce, how active these enzymes are, and what tasks they perform in the organism. This new category of behavioral genetics, which takes advantage of gene chips and association study models, holds great potential for development. It also holds great potential for headlines. When the media report time and again that someone has found the gene "for" something – the "smoker's gene," the "infidelity gene," the "gene for bad driving" – the underlying study is sure to hail from molecular research.

IN THE EARLY days, this sort of molecular work didn't garner a positive press. In 1993, when the American behavioral geneticist, Dean Hamer, pointed out that a particular region on the X chromosome appeared to be associated with a predisposition to homosexuality in men, it stirred up a brouhaha. Hamer did not claim he had found a predisposition that would be expressed every time it appeared, or one that is found in all homosexual men, just a predisposition that proved to be statistically over-represented in the relatively small sample of gay men he had studied. His first group of subjects comprised forty pairs of brothers, some of whom were homosexual, and a later study of a new group again found the same region: Xq28.

Various gay organizations went into celebratory mode, because Hamer's results set the stage for ironclad arguments for equal rights and antidiscrimination measures: since sexual orientation is a part of your biology, others could hardly accuse you of perversion and being unnatural. Outside gay circles, the finding was less popular, particularly among certain religious camps that claimed to excel at converting homosexuals to heterosexuality. In their view, homosexuality was (and is) a behavior choice. Beyond religious groups, there was also resistance. Some shouted that, in the name of political correctness, scientists should not even be allowed to research these sorts of topics. What if a "gay gene" did indeed exist? Would it inspire expectant mothers to ask for prenatal genetic testing, and would some of those to mothers possibly decide to abort a "gay" fetus? Even fellow scientists attacked Hamer for, among other reasons, wanting to reduce complicated psychosocial links to simple biology. "This is the most solid work our group has ever done, but we were forced to abandon it at the time, because it was impossible to get financial support," Hamer said to me, when I spoke to him much later about the events. "The debate was incredibly shrill and went far beyond the science to pure ideology about gays."

Increasingly shrill voices were heard later that year when the Dutch scientist Han Brunner found the "aggression gene." Brunner had studied a Dutch family in which, over five generations, a conspicuous number of men were violent, criminal, and exhibited subnormal intelligence, and had discovered they all carried defective versions of the MAOA gene. This gene codes for the enzyme *monoamine oxidase*, which breaks down a number of neurotransmitters such as epinephrine (adrenaline), norephinephrine (noradrenaline), and serotonin. In these aggressive men, however, the defective gene blocked production of the enzyme and an overflow of the transmitter substances became available in the brain. The discovery forced the public to consider whether aggression and other impulsive behavior could be traced back to MAOA variations. Some commentators fretted that violent criminals would demand gene tests and use "my genes did it" as a defense; others predicted the screening of problem children and subsequent genetic stigmatization.

Because of these early controversies, the public was happier to get the news that a study had found a genetic key as to why some people have a particular need to throw themselves into experiences and activities that provide a thrill. In two articles published in the prestigious journal *Nature Genetics* in 1996, American and Israeli researchers independently uncovered a connection between a particular variant of the gene for the dopamine D4 receptor and a thrill-seeking personality. There was even a physiological logic to the link.

DRD4, as the receptor is called, is a protein located on the surface of certain brain cells, especially in the areas of the brain that process emotion – the limbic system. Here, the receptor captures and binds the neurotransmitter dopamine and passes on signals to its cell. How powerful the signal is depends on the structure of the receptor. If you have a particularly long version of the DRD4 gene, you get a receptor that binds more weakly to dopamine and provides a poorer signal. The researchers' hypothesis was that people who are blessed with the long variant of DRD4 must do something extra special to get their shot of dopamine and its accompanying glow of well-being, compared to the rest of us. While the possessor of a short DRD4 can satisfy his lust for adventure by watching a mountain-climbing flick at the local cinema, the person with a long DRD4 has to put on her boots and trudge up Mount Everest. When the scientists drilled into the data, they found that the long DRD4 variant did not guarantee a personality with a need for thrill-seeking and danger, but it was an important component. That hypothesis was supported when, a year later, a study looking at a group of Japanese subjects reached the same result.

For the first time, behavioral genetics had pinpointed a single gene with a specific effect, and things began to move. In early 1996, a team from the University of Würzburg, led by Klaus-Peter Lesch, announced they had achieved a breakthrough in understanding the personality trait of neuroticism. Roughly speaking, neuroticism involves a tendency toward anxiety, to brooding over problems, and to getting hung up on the negative aspects of life. Lesch's team found a connection between a subject's high neuroticism score and a particular variant of the serotonin transporter gene, known as SERT. The gene sits embedded in the surface of brain cells, capturing the neurotransmitter serotonin and sending it back into the cell after it has been released to trigger a signal. One component of the SERT gene is a regulator that determines how much of the protein used to transport serotonin is produced. Akin to DRD4, this regulator is found in shorter and longer variants. In this case, the short version has the effect that less of the transporter protein is manufactured. In Lesch's study, the subjects with a high neuroticism score typically had one or two copies of the short variant, whereas the less neurotic far more frequently had two copies of the long one.

You can see a recurring pattern. Again and again, the genes that behavioral genetics happens upon turn out to be genes affecting the chemistry of the brain, especially those that control the breakdown of neurotransmitters.

At the same time, these genes are typically promiscuous – they are likely to be significant for many different traits. DRD4 is a good example. According to a study carried out by researchers from the Hebrew University in Jerusalem, not only does it help define our appetite for everything new and exciting, it also plays a role in fine-tuning our sexual behavior. Different variants of DRD4 apparently influence such things as our level of sexual desire, how excited we become during the act, and how well we perform.[13] Something similar holds true for MAOA. Since its first appearance with a connection to aggression, variants of MAOA have been implicated in "social sensitivity," as well as compulsive gambling and psychiatric conditions such as hyperactivity and obsessive-compulsive disorder.

———

BEYONDTHEIR PRESENCE in the brain and their broad repertoire of effects, there is another common feature of gene variants: their effects are rather small. Molecular studies do not throw off the strapping, healthy numbers of heritability familiar from twin studies: thirty percent, fifty percent, eighty percent. When, for example, scientists look at the extent to which SERT determines neurotic temperament, the short SERT variant can explain eight percent of the overall genetic effect on variability in anxiety. A measly eight percent, you think, but in this business, this is actually very large.

Compare that eight percent to the findings for intelligence, for instance. For over twenty years, Robert Plomin, a professor of child psychiatry at King's College in London, has searched high and low for IQ genes and found pretty much *nothing*. Not a thing. In his latest study, Plomin and his colleagues have studied all six thousand or so British twins born between 1994 and 1996, whom they have followed through the years and measured this way and that. The researchers used gene chips to compare a half million SNP markers in high IQ and average IQ children. The only difference that emerged from these rigorous statistical tests was a marker in a gene for which no one knows the function – a marker that, according to calculations, can provide at most an extra half point of IQ.

"It is, of course, a little disappointing," the American Plomin says, in acquired British understatement, when I visit his lab in the London district of Brixton. Despite the disappointment, however, he remains in good spirits, and had started to gather support for an international study of ten thousand children across Europe. His expectations were low. As Plomin put it: "We may be able to explain between three and five percent of the variance in intelligence in the next five years, and it may involve hundreds of genes."

So, what's gone wrong? It may be, as Dean Hamer pointed out in a commentary in *Science* magazine several years ago, that researchers need to rethink their whole approach to behavioral genetics. They need, says Hamer, to reject the classic model, which assumes a simple linear relationship between genes and behavior; that a particular gene provides direct access to a particular form of behavior. In other words, genes do not code for behavior, they code for proteins, pure and simple. There is a black box wedged between the two, in which networks of genes work together with environmental factors and with the brain's physiological development and function to create what we call "behavior." It is this box researchers must try to prise open.

And they're working on it. Recent studies have focused on how genes interact with specific environmental characteristics, using the model of disease research. They look for genes that predispose or provide sensitivity to a disease and then look for aspects of lifestyle or surroundings that precipitate this predisposition or sensitivity.

One couple has become known for leading the way in this approach to behavioral genetics: married psychologists Avshalom Caspi and Terrie Moffitt, who are affiliated with King's College, London and Duke University in North Carolina. Just after the turn of the millennium, their studies took psychiatry by storm.

Caspi and Moffitt gained access to a goldmine, namely, data from almost a thousand New Zealand men whose psychological and social development had been charted for more than twenty years, from the time they were in nursery school. The researchers screened the adults for "antisocial" behavior, defined as problems with aggression and violent criminality, among other things. The next step was to conduct gene tests to investigate whether there was a connection between antisocial behavior and variants of MAOA.

There wasn't. You could not predict from an individual's MAOA genes alone whether or not he would have problems. But another factor did come into play: men who had inherited the MAOA variant with low enzyme activity from both their parents had an extremely high probability for developing problems if, at the same time, they were subject to a bad early life experience, with abuse in the home. It wasn't the gene itself but an explosive cocktail of biological inheritance and childhood experience that contributed to later aggression.

A year later, Caspi and Moffitt conducted a similar study to look at the SERT variants and depression. Once again, the connection between gene variant and psychological condition proved to depend on upbringing. The researchers counted up how many particularly stressful life events the individual had experienced in the previous five years and asked whether he had experienced a clinical depression during that same period. The frequency of depression depended on which variants of SERT the men had. Of those who had two copies of the short variant and had been through more than four major stressful events in the five previous years, almost half had experienced clinical depression. By contrast, for men with the same number of stressful events but two copies of the long SERT variant, only one in four had been clinically depressed.

On top of that, there was a clear effect from childhood abuse. Among the men with two short SERT variants, the risk of being hit by depression as an adult increased powerfully – up to two in three – if he also had been abused as a child. Among those with two long SERT variants, child abuse made no difference in the risk for later depression. The short SERT variant provided an increased sensitivity to depression, and that sensitivity was increased further if the person had a rough childhood.

I'M THINKING ABOUT depression on the train from Washington, DC, to Richmond, Virginia. I'm on my way to meet Kenneth Kendler, who has been a psychiatric epidemiologist and a professor at Virginia Commonwealth University since 1983. I've met him before. Many years ago, he graciously granted me an interview at a fancy Copenhagen restaurant, where I enjoyed a big, very rare steak, while Kendler stuck to his vegetarian diet.

The train trip from the nation's capital is typical Amtrak. It takes a miserably long time; the train moves at a snail's pace, and the temperature inside is freezing cold, to counteract the sticky humidity of June. A number of passengers are wrapped in jackets and cardigans, a rebuke to the summer that ambles by outside. The view from the window is unvaryingly green, so there is not much to do but read. For entertainment, I can choose from *Physiological Medicine*, *JAMA* – the *Journal of the American Medical Association*, *General Archives of Psychiatry*, and the *American Journal of Psychiatry*. I leaf through each in turn, and then my gaze falls randomly on the conclusion at the bottom of the first page of an article in the last journal: "Major depression is a familial disorder, and its familiality mostly or entirely results from genetic influences."

I scan further up the page. It says that a number of studies of depressive families show that the condition's heritability is around forty percent. They could have aptly studied my family. Just my first-degree relations would have provided reams of quality data. For instance, there is an unbroken line from my maternal great-grandfather to me, all of whom were hit by depression. Great-grandpa Marinus Hansen – whom I only know from family stories and his glum, grey photo portrait – put a bullet in his forehead; his daughter, my maternal grandmother, was almost subjected to a lobotomy. It was a ghastly but heart-warming story, which I loved to be told as a child.

"Mommy, mommy, tell me about the time your mother almost got lobotomized," I would beg, and my mother would happily relate the tale. It was, after all, something she'd experienced as a teenager. At that time, my grandmother, who had worked insanely hard for her whole life, broke down into a deep depression. So deep, she had to be hospitalized, and for months just lay in her bed. No one could do anything. Not that they didn't try. The chief surgeon was particularly generous with electro-shock therapy, but nothing helped. Mrs. Frank was too sick to act for herself, and finally the doctors recommended that the family try one last procedure: psychosurgery.

In the 1950s, medical reports indicated that a lobotomy — where they rummage around with a thin metal spatula in the patient's frontal lobes, cutting random nerve paths — could relieve very severe depression or, at any rate, bring patients to a state in which they could be sent home. The chief surgeon suggested they give it a go. My maternal grandfather, who knew nothing about psychosurgery or medical treatments in general, was in favor of signing the papers.

It was here I always felt goosepimples, no matter how many times I'd heard the story. *Think if they'd actually done it.* But the family doctor put his foot down — this would not take place under any circumstances. Instead, he found a psychiatrist, a specialist who lived far away but to whom my grandmother was driven once a week for two years. It helped. I remember the grandmother who emerged only vaguely — she died before I started primary school — but my recollection is of a loving and strong woman. A woman who, after fifty years, no longer allowed herself to be bossed around or to jump through hoops for anybody.

Presumably, my grandmother passed the tendency toward melancholy on to her daughter, my mother, who fell to pieces over her mother's illness. She dropped out of secondary school, where everything had been going so smoothly. She spent a couple of years alone with her father. She pretty much stopped eating, eventually developing a metabolic disorder. She never used the word "depression," but when she talked about her early youth, it had the unmistakable sounds of one long clinical episode. Later, the illness returned with such force that I don't remember the final years of her life without a shadow coloring everything.

Suddenly, I'm sitting here, half way to Richmond, shivering like a newly shorn lamb in the train's artificial deep freeze, feeling lucky.

Really lucky. In comparison to the generations before me, I have had it incredibly easy. My time as a teenager was not an outright catastrophe, just garden-variety terrible. As to depression, I only have three light episodes to boast about. Moreover, I live in a time not only when depression can be treated quite effectively but when you can also begin to delve into and understand this strange and burdensome heritage, a heritage people once wanted nothing more than to ignore.

IT'S RAINING IN Richmond. A constant drizzle that is not only sad in itself but makes the surroundings grim. Or, rather, grimmer; Virginia Commonwealth University is located in a pretty hideous part of town. The university buildings are bland and new, and surrounded by timeless decay; here and there are old two-story brick buildings in disrepair, and street-level shops that have gone bankrupt and been boarded up. The neighborhood, the weather, it's enough to get you down.

Until you turn the corner to West Avenue, when the scene is transformed into an all-American small town, with clapboard houses, overgrown gardens, and tree-lined streets. I glimpse Kenneth Kendler through the large front window of his house. When I get inside, he can hardly conceal his excitement.

"You must have seen today's *New York Times*?"

I show him the front page of my *Washington Post*.

"Yes, it's the same story, and I already have twenty e-mails about it." We stare at the headline: "Report on Gene for Depression Is Now Faulted."

"Excellent, excellent," says Kendler, mostly to himself, "let's go inside."

From my chair tucked into the front window's bay, I notice all the signs that a cultured family lives here. The rooms are decorated with antique furniture and aged, original prints from India. Above the dining room table hangs a large, color photograph of a young woman who clearly belongs to the present but is posed as a Renaissance figure from seventeenth-century Holland. It is a strangely brooding portrait.

"My daughter," explains Kendler. "She's an art student. It's a self-portrait."

More conventional pictures of her and his two grown sons fill the landing, where I also notice a photo of the parents, when they were young. It seems to be from the '70s, but Kendler looks nearly the same today. The only difference is that his hair and beard are no longer black but grey. Just like the last time we met, he reminds me of a thoughtful rabbi. Perhaps, it's the gaze behind his round, owlish glasses or the way he speaks with a low, penetrating voice and an almost exaggeratedly clear diction. Kendler is a very calm, polite man. The strongest expression ever to pass through his lips is "bull-shit," and that happens only very rarely.

"I grew up in a Jewish community on Long Island, outside New York," he says at one point. "It was all very traditional until I moved to California as a teenager, where there was a different pace with drugs and …"

Unfortunately, he doesn't go into detail. We are interrupted when a very reserved tabby cat with pale green eyes comes slinking in. The cat glances at her guest before creeping up the last few inches to sniff my outstretched hand.

"She is very particular and doesn't take to just anybody," remarks Kendler. Today it's not a problem, because I'm very good with animals. They love me, I explain, and I speak to the cat in the manner I use with my black tomcat at home. "Such a sweet little pussycat, come here." The cat draws nearer, but when I try to scratch behind her ear, she bites aggressively at my hand and plants a single cuspid into the soft flesh between my thumb and forefinger. Then, she slowly curls into a ball on the floor and fixes me with a malicious stare.

Meanwhile, Kendler returns to that day's big news, in which he is cited several times. The story involves Avshalom Caspi and Terrie Moffitt's famous Dunedin study that showed the gene for SERT increases the risk of depression if the person experienced traumas and abuse as a child. Now, however, a group of psychiatrists and statisticians have produced an analysis of fourteen other studies using the same type of data and, when all this data is put together, they cannot see any connection between SERT and depression.

"I should think that would take the wind out of the sails for Caspi's studies," says Kendler. He hurries to add that Avshalom's father was his Hebrew teacher in school and that, no doubt, the younger Caspi is a fine fellow. The two iconic studies of the young men of Dunedin and the interplay of the genes for SERT and MAOA in their childhood remain, in their way, an excellent starting point. The studies have simply, in Kendler's opinion, acquired influence far beyond what they deserve.

"My explanation is that these two findings represent an ideal answer to the old question of nature versus nurture. You can almost say it is 'feel good' research. The results provide a sense that everyone is a winner. It is very intuitively appealing, and when it was published in 2003, everyone who had done studies with the SERT gene immediately went to their own data to see whether they could find the same effect if they asked the research subjects about their childhood."

"And …?"

"Some could. Others couldn't. And the new meta-analysis takes them all into consideration."

Meta-analyses. These are statistical exercises based on testing a connection that is shown in some studies but not others. They work by merging all the studies together and treating them as one. Meta-analyses are often thrown on the table as a sort of trump card in research. Smack! Take that. We have more data than you, so we're right.

Nevertheless, many people in the field have a hard time accepting this as the end of the entanglement of SERT with depression. After all, the "Woody Allen gene," as it has been called, has repeatedly been linked to a sensitivity to stress – the short version of the gene makes you worse at dealing with stressful situations and events. In 2007, a team of British researchers reported that a timid temperament in children linked up nicely with the short version of SERT, and that same year another group saw a connection between the variant and suicide. At the same time, stress is frequently indicated as a triggering factor for depression. If I remember correctly, Kenneth Kendler and his group were among the first to show, back in the 1990s, that people react differently to influences associated to depression. That is, sensitivity is not uniform.

"Mind you," he says very evenly, "I, too, believe the general phenomenon described by Caspi exists. There is just not enough effect in each gene to see – statistically – interplay with environmental factors such as upbringing. But it is also a fact that psychiatry is an immature science. People get wildly enthusiastic about one idea and then another – it looks almost like infatuation. Psychiatry is influenced by fashion, and it was fashionable to study the environment and individual genes in the way they did in the Dunedin studies. My hope is that people will cool down and look more soberly at the data."

I mention that my personal interest in depression comes from my family background, and I am curious to know how genes and upbringing interact.

"Yes, of course," he says sympathetically, like a funeral director. "Let me tell you something. My own starting point in psychiatry was a deep interest in schizophrenia. But schizophrenia research has gradually moved in the direction of genetics, and one of the reasons I later worked so much on depression is that schizophrenia is simply *too* hereditary – so genetically determined that it is too difficult to study many of the interesting questions around the significance of external influences. In depression, on the other hand, there are stronger environmental variables. We know, as you yourself say, that stressful events such as death and divorce are clearly a factor in triggering the disease."

Just a minute. Is a professor of psychiatry and genetics actually telling me that there is too much focus on genes?

"I can see a sort of fixation on genes, yes."

But is it unhealthy?

"I don't believe for a moment that we even *can* find genes that solely and in themselves have a major influence on whether you develop a psychological illness or not. In other words, we won't see a parallel between depression or anxiety and Alzheimer's in which a single gene – ApoE – actually says it all."

That, I think, *is* interesting, and perhaps a bit peculiar to hear from a man who was recently appointed the head of a gargantuan genetics project. Together with researchers from Oxford University and Fudan University in Shanghai, Kendler is gathering data from six thousand Chinese women with repeated depressions in order to use gene chips to study a half million genetic markers, much like Robert Plomin's protocol for investigating the heritability of intelligence. So, the man who apparently does not believe in the discovery of decisive genes is doing a traditional genetic association study to find new genes.

"Correct. But note that we are not just studying the genetics of the women; we're also assessing their environment very thoroughly." Kendler breathes deeply and looks over his glasses.

"Even though the genes we are finding are not in themselves decisive – rather, each plays only a quite small role – they point in the direction of some concrete biological pathways. They give us a key to open the door to the underlying mechanisms for illness about which we are still fundamentally ignorant. We believe, however, that neurotransmitters such as serotonin are involved in depression, because the drugs that inhibit symptoms act on serotonin. But, in reality, we are very far from knowing what is going on."

The serotonin hypothesis is the one everybody has heard of. "You're depressed? Well, then, you have too little serotonin," they say, or, "Your serotonin balance is off" – whatever that means. A few years ago, Jeffrey Lacasse of Arizona State University and Jonathan Leo of Lincoln Memorial University pointed out in the periodical *PLoS Medicine* that this is a message we particularly hear from the pharmaceutical industry. The duo had studied advertisements for SSRI drugs and found that they consistently play up the claim that well-known brand names such as Zoloft, Paxil, Prozac, and Lexapro normalize the brain's serotonin balance. But there is no scientifically established "balance" for serotonin, Lacasse and Leo assert. In other words, no one has measured how much serotonin there *should* be in various places in the brain to keep someone free of depression. Or, as the British psychiatrist David Healy, a specialist in antidepressive remedies, once put it to me in an interview: "The theory that serotonin levels are the cause of depression is as well founded as the theory of masturbation causing insanity."

But then we are back to how you are supposed to study environmental influences.

"We've only just begun to look at what happens during the development from child to adult," Kendler says. "I bet you there is something very interesting there. There is a conception that our genes are immutable – we inherit them and that's that, nothing happens after that. But that is far from the truth. You can see that the genome is very dynamic over time in the way that genes associated with certain traits at the age of seven do not show the same association if you check at the age of sixteen. Some kind of rearrangement of genetic effects occurs at puberty."

Gender differences are another area.

"Take something like the frequency of depression: it is the same for girls and boys until the age of fifteen. Then, women take the lead for the rest of life. Adult women suffer from clinical depression twice as frequently as men do."

Depression is heritable – on the average, in the vicinity of forty percent – but if you focus on the hereditability, it looks as though genetic factors play a larger role in women than in men. That is, the heritability is greater in women.

"You can ask whether the same genes always give the same effects in men and women and, for depression, we can say that they don't."

But as soon as you get into gender differences, you inevitably also think of hormones. Just take premenstrual syndrome, the recurring monthly symptoms that, for some women, approach a depressive mood. So, couldn't it be estrogen that makes the difference?

"Hormones are certainly a part of the story," says Kendler affably, but with a hint of indulgence in his voice. "But doesn't that also mean that society should treat women differently as soon as they are sexually mature?"

I don't quite get what he's aiming at.

"Wouldn't all the hullabaloo around body image and self-esteem play into this?"

I really couldn't say.

"For example, studies show that early menstruation is a very bad thing for girls who are in mixed gender classes. Early maturity makes for early older boyfriends, an early sexual debut, more instances of drug abuse, and mental illnesses over a lifetime. If the same girl, on the other hand, goes to an all-girl's school, you don't see the same connections." Kendler enunciates his next sentence very slowly and carefully. "This is very much due to the *reaction* of those around her to something biological. There is nothing so inviting for an older boy as a physically well-developed thirteen-year-old girl who is mentally too young to defend herself."

I ask whether this might be genetic to men, and Kendler smiles slightly wearily. "I'll dodge that question," he says.

On the other hand, there's another promising research area that he's happy to discuss: the interactions between genes and environment. "We have to understand the field of probability in which genes increase the risk of developing a psychological condition if we are subjected to certain stress factors," he explains. "One of the challenges is what I call genetic control of environmental exposure. It is particularly important in conditions such as anxiety and depression."

KENDLER IS CONVINCED that one way in which genes work in relation to psychological illness is by changing the environment around us. How in the world would that occur? In 1997, he demonstrated through a twin study that the psychological concept of a *social network,* which details the interpersonal ties and support systems available to an individual, is not only socially determined but also highly heritable. This finding encouraged him to speculate that partially genetic traits, such as temperament, might cause people to build their own social environment.

"As we go through life, we actively create our own environment through the way we interact with other people. And the quality of this environment, in turn, affects our psychological state. There are circles that run into each other – the genes that make us inclined to develop depression are also the ones that make us more difficult to be married to or work with."

When people say "environment," most of us think of things that just happen. The music of chance, so to speak.

"Shit happens, they also say. Of course, chance and bad luck are part of the game, but if you look closely, many negative life events have to do with human relationships – especially when it comes to couples. We're not just victims, we're accomplices. We have done long-term studies of neurotics that illustrate my point. By following a group of people and their lives over many years, we have shown that if you measure how neurotic – that is, how nervous, sensitive, and so on – a person is, you can *predict* future life events with respect to personal relationships and the ability to have a social network."

The more neurotic, the more problems, the smaller the network, and the greater the risk for depression and anxiety.

"When we began to publish this stuff, we got a lot of resistance from colleagues in psychiatry. It just couldn't be true! But, funnily enough, it was an entirely natural thought process for evolutionary theorists. Everywhere in nature, genes work by changing the environment around them. Genetically different weaverbirds weave different nests and thus attract different types of females. And certain genes in a cold virus try to irritate the mucous membranes in our noses to make us sneeze and thus spread virus genes."

It reminds me of a column that ran on the *New York Times* website under the headline "Mugged by Our Genes?" The writers were probably inspired by the Florida State University criminologist, Kevin Beaver, who had come to the conclusion that being the victim of a violent street mugging was actually partially heritable. *What?* Isn't being the victim of a crime just a question of bad luck? But after poring over data from a major twin study that followed young Americans over part of the 1990s, Beaver calculated that almost half of the variance in relation to whether they were mugged had to be ascribed to genetic factors.

Upon closer consideration, this is not as absurd as it sounds. Beaver's thesis is that genetic effects work indirectly. They help shape behavior that increases the probability that a person's path will cross that of a mugger – perhaps because he seeks environments where criminals are more likely to hang out.

"This effect presumably explains why the heritability of mental traits seems to increase with age," says Kendler. "After childhood, we have more freedom to go where we want to, and genetics plays a role in our preferences."

Isn't it like a snake eating its own tail? And if genes also help determine our environment, doesn't this suddenly leave less room for free will?

Kendler leans back and chuckles mischievously.

"I've been thinking about all that free will stuff, and I believe you are falling into a classic trap," he replies. "People always think that genetics and inheritance are about determinism, while environmental influences aren't. But think about it: if the protein content of your diet in your first three years of life influences your brain and, thus, your intellectual potential and your personality, isn't that also a loss of free will?"

His question, of course, is rhetorical.

"If we accept that our behavior comes from the brain and that the brain is a biological system in which there are causes and effects, then this is, in reality, the only problem we need to concern ourselves with. Whether depression is sixty percent or ninety percent heritable, the behavior comes from the brain. There are biological limitations. The extent to which they are created by genes or environment, or both is, in fact, immaterial."

The conversation has taken a somewhat dispiriting turn. It's difficult not to get a little discouraged when the topic of free will comes up, because nothing ever seems to come of it. Purely subjectively, we all *feel* as though we have free will. We could, after all, say "no" to that third glass of wine during happy hour, just as we could keep within the speed limit on the highway. It's just a matter of choice, right?

When you think about it, though, it's clear we are just fooling ourselves. Obviously, no one is completely free in the classic sense of being able to choose anything, in any situation. Such freedom does not exist. We all find ourselves in cages limited by our being, our history, and our experience. But the challenge is whether there are any tools for expanding the cage. Could possessing all this genetic knowledge change the way we think about ourselves? Do we ultimately gain a little more free will by knowing our biological limits?

"That is, of course, a very interesting question," says Kendler, who then falls silent. I can't quite judge whether there is a slight tinge of sarcasm in his tone but, fortunately, he continues in an obviously serious vein.

"It also says a lot about our image of ourselves and the peculiarity of Western culture, which has to do with the desire to relate biology to responsibility. But whether we can unite knowledge of biological causality with the concept of free will is not a scientific question at all. It is a philosophical problem. As soon as biology gets into the debate, it actually has to do with the omnipresent idea that we can shape ourselves, that we can make a stressed-out, maladjusted person completely calm if he just takes a course of transcendental meditation. Or that a navel-gazing, introverted person can be transformed into an exuberant extrovert if he gets a little therapy and engages in positive thinking."

Kendler shakes his head slightly.

"But the idea that we are all equally sensitive is unrealistic. Human experience is a subtle, nuanced system in which we come to the world with some very substantial dispositions with respect to intellectual capacity and personality traits. Dispositions that we can only shape to a certain degree."

He remains quiet but rapidly brightens up.

"Let me give you an example where the evidence is quite good: a child with a moderate case of ADHD, and behavioral problems. You know, a child who shows highly impulsive behavior, can't sit still, and so on."

I try to imagine this hypothetical brat. I raise a hazy picture of Niels, an acquaintance from my early childhood. He was a particularly restless boy, who bit and kicked and threw scissors at people until he was finally expelled from nursery school. A seemingly archetypal case of attention deficit hyperactivity disorder. Later, I heard, he became an experimental psychiatrist.

"Some parents can shape these children in a positive direction by being a sort of buffer for their propensities. These parents, in fact, reduce the heritability. But other parents exacerbate the same conduct. But people cannot stand the complexity of a nuanced problem. I notice this when I'm out speaking about this subject in public – there is a huge desire to be able to lean back and say, 'It's my genes, it's not my fault.'"

The fact of the matter is that, by knowing more about inheritance, we can learn more about how we shape an environment – or engage in an "intervention," as therapeutic lingo would have it – to nudge our genetic background in a desirable direction.

"That is partly correct," says Kendler.

So, what can you do about a person with an early tendency toward depression, just, you know, hypothetically speaking?

"Relatively little, I would think, but I don't know the intervention literature very well."

Don't they at least know what type of environment or circumstances depressive people typically seek out?

"They know a little, in an indirect way. We've done twin studies that look at neuroticism as a trait closely associated with a vulnerability to depression. Neuroticism increases your risk for having a lower degree of social support, for having a poor network, and generally meeting with negative events. This indicates, in other words, that the component of the genes that increases your risk of depression has the effect of giving you a tendency to have difficulties with human relationships."

But, he stresses, you can't provide a gene test for that.

"We find ourselves at a stage where there are some research findings – interesting research findings, at that – but these don't yet constitute enough of a basis to form a conclusion."

People want conclusions, I argue, dumping all my expectations for a bit of progress on my host. Doesn't he think that genetics, now that it is accessible to everyone, will appeal to the human need for prediction? We want so much to optimize this one little life we have. We want so much to have some guidelines and directives that make sense and that can, in some way, guarantee that we exploit our potential to the max. Is he really saying that things will never be certain, nothing will ever be on solid, scientific ground?

"I say," Kendler replies slowly, "that it is difficult to know."

I GET NO more answers. Sue Kendler has put homemade brownies and some fruit on the table, and I take the seat below the daughter's self-portrait while the crazy cat with her grape-like eyes glowers from the nearby fireplace. We chat about this and that, until I think it's time to get back to my hotel bed in Washington, if I could just call for a taxi to take me to the train. The couple looks shocked.

"Train?There's no train now, it's almost nine o'clock," Sue says, explaining that the train traffic in and out of Richmond stops around mid-afternoon. My palms go clammy. I don't get it. Washington is the nation's capital, and the train ride takes just over an hour – how is it possible that you can't leave this small town after mid-afternoon?

"This isn't Europe," Kendler says with a smile. But Sue immediately invites me to stay overnight. Two of the children are home on vacation at the moment, taking up the guest rooms, but they have a comfy sofa I can sleep on.

"Please, stay and take the train back tomorrow."

I feel shy and awkward at their invitation, incapable of accepting. I mumble something about finding a hotel, and it ends with Kendler driving me to the local Holiday Inn, which is nearby and within my discount budget. It is a sad place. One of those motels where the buildings have two floors, and the doors all open onto a concrete balcony. Everything is brick and avocado-green wood, reminding me of a cheap university dormitory, circa 1970.

I'm pretty much the only guest and wind up sitting in my room drinking tap water – which tastes of chlorine – and wondering why in the world I did not accept the Kendlers' hospitality. It's crazy to sit here completely alone in a semi-dark motel room with tasteless decorations on the walls and watch bad television. Instead, I could have got to know some intriguing people better, and I might have been able to learn something about the complicated connections between the human mind and DNA. After all, that is what I traveled all the way from Europe for.

This may be precisely the type of situation Kendler was talking about. This is not unfortunate circumstances but *me*, the neurotic, actively seeking out situations and shaping my environment. I am isolating myself and boring myself – fertilizing the soil of melancholy and self-pity. Depression, if I'm fortunate. Maybe it's the chlorinated water and the quiet, but I now see myself as a behavioral geneticist might, the result of a long series of self-selected circumstances.

Even as a snotty-nosed kid, I would trip myself up socially. For instance, I loved to go to my two cousins' house. They lived far away, were a few years older than me, and were, therefore, endlessly attractive. But what did I do when I finally got there and my cousins had been told to look after me?

I would let them know straight away that I only wanted to play chess, a pastime they did not care for, and they would respond by locking me in the guest toilet and telling me to stay there until they gave me permission to leave. And there I would sit until some merciful adult happened by and let me out.

Ever since, I have regularly felt banished to the loo, metaphorically speaking, but have never really considered whether I might have put myself there. Is that my genome sneaking out in subtle ways? My conversation with Kendler about the weird tango between immovable DNA and our dynamic psyches has given me a new, intimate understanding of myself.

FAR FROM THE Holiday Inn and safely back in Washington, I think again of Dean Hamer, the discoverer of the "gay gene." Or, more precisely, I think of something Hamer said several years ago about what will be required for behavioral genetics to match the complexity of twin studies.

"Geneticists have to start working together with brain researchers who are using more and more sophisticated scanning techniques to analyze the functions of the brain," he said. "Instead of trying to make a leap directly from gene to behavior, we need to take small steps. From gene to biochemistry, from biochemistry to what happens in the brain, from the brain to behavior."

One of the people taking these steps is Daniel Weinberger, of the US National Institutes of Health. For all practical purposes, he has invented the field of *imaging genetics*. As that name implies, Weinberger aims to create pictures of the genetics under investigation. He couples advanced forms of brain scanning, which capture what is going on in the living, thinking, and feeling brain, with knowledge of which gene variants exist in that brain. The approach allows scientists to look beyond behavior and directly at the processes occurring in the brain.

Take, for instance, imaging studies of our old friend SERT. After debating for years how – or whether – the gene was involved in shaping susceptibility to depression and neuroticism, Weinberger decided in 2002 to put it under the microscope. He got hold of research subjects who had either two copies of the long SERT variant or two copies of the "sensitive" short SERT variant. Everyone took a trip through an MRI scanner, which was set up as a cinema for the occasion. The volunteers didn't have to do anything but lie completely still while they were shown a series of pictures of unknown faces with different emotional expressions, from the funny, to the fearful, to the furious. The imaging showed a distinct difference in the activity in the amygdala. This minute brain area, which is involved in signaling negative emotions such as fear, disgust, and loathing, lit up when researchers showed pictures of fearful or angry faces.

The "sensitive" ones – those with two short SERT variants – reacted with a significantly higher amygdala activity than the others. Presto! Here was something you could put your finger on. Instead of a vague statistical connection between a gene variant and a complex condition such as depression, Weinberger and his colleagues had identified a concrete biological mechanism arising from a tiny genetic difference. It was a first glimpse into the mysterious black box that lies between genes and behavior.

To meet Weinberger, I've rented a car, but as soon as I arrive at the NIH campus in suburban Maryland I realize this was a dumb idea. I feel as though I've arranged a meeting at a military camp, not a research institute. The safety inspection rivals anything I've experienced at an airport; if you are unlucky enough to be in a car, as I am, you are summarily escorted out of it so that a bomb detection dog – a friendly yellow Labrador – can check out every inch of the interior. While the dog's sensitive nose conscientiously explores the garbage I've thrown on the floor of the passenger side and lingers over a sock that has somehow landed on the back seat, I am scanned for hazardous substances in a small glass cage. My passport data is typed into a database and, when I complain, the security officer at the computer flatly replies, "These are the safety precautions since September 11." He then looks me up and down before adding "ma'am," which makes me furious.

"You should probably calm down," advises another guard. They are an interesting crew: all very dark-skinned, and boasting identical African accents and handguns – presumably loaded. I get the absurd impression that an African competitor of the Blackwater security firm has won the concession to guard NIH's scientists. I manage not to blurt the idea out, and one of the kind gentlemen sends me in the direction of Building 10, a vast, burgeoning structure on campus that proves to be a modern labyrinth.

"Weinberger? Is he a patient here?" asks a sullen woman at reception and, when I explain that he is a professor, the head of a research group in the neurogenetics department, she frowns and looks disappointed.

"Then I really can't help you."

I ask at five different desks and get sent in five different directions – all wrong. Not until a local woman takes pity on me and takes me by the hand do I make any progress.

"Just calm down," she says in a tone that indicates that I'm behaving like a spoiled child. "It took me a month to be able to find my office without getting lost."

With her experienced guidance, we find it. An insignificant door in a distant corner opens onto Daniel Weinberger's department and shuts out the bustling lab coats and impatient hospital patients. A secretary welcomes me and gestures me to a comfortable chair. On the wall facing me are pictures from international conferences in which Weinberger's research group has participated, and which, interestingly enough, all seem to have taken place at ski resorts or on the beach. One looks to be the Alps, and several – to judge from the apparel and the many tiny umbrellas hugging the lips of long drinks – must have been in Hawaii.

Weinberger is grinning widely in all the pictures. And when he comes out to greet me, he looks like a man who doesn't go out of his way to avoid a party. His face is slightly worn and the bags under his eyes appreciable, his voice sounds like no small amount of cigarette smoke has passed through his throat, and his accent is the broadest Brooklyn. He belongs to that part of humanity that you immediately want to drink a beer with.

"Would you like a soda?" he asks. "It's diet."

While he procures the yellow drink, I glance around his office: cluttered, with the usual family photos and, on the windowledge, a row of volumes squeezed between bookends shaped like the two hemispheres of the brain. I wonder why scientists almost always have a weakness for that kind of junk. They always seem to have small monstrosities that you'd never find in the office of an architect or an investment banker.

"What can I do you for?"

Weinberger's manner is about as far from the reserved Kenneth Kendler as you could get, but the two have the same point of departure – that is, they are psychiatrists with an interest in schizophrenia. In the 1980s, Weinberger studied dozens of twins of which one was schizophrenic and the other was healthy, because he hoped it would yield some insight into what in the disease was genetic and what wasn't. He was particularly interested in whether, beneath the illness itself and its symptoms, there were any ways of thinking or functioning that those with the disease shared with the healthy. Things were going swimmingly until a powerful new model arrived unexpectedly.

"I'll never forget it," Weinberger says hoarsely. "It was 1992, and I was at a meeting with Harold Varmus, who headed up NIH at that time. 'Hello!' he said to all of us, sitting there with our big egos.

'You've been researching schizophrenia for twenty years without finding anything at all of significance. From now on, you're going to be doing genetics, folks. The mapping of the human genome is going full steam ahead, and the project will identify genes that are involved in disease and every kind of human conduct. If you don't study genes, you'll very soon be like a flock of dinosaurs caught in the age of mammals'."

Weinberger's laugh is unvarnished and hearty. Liberating.

"I knew it at once – the man was right. Here I was studying the symptoms of schizophrenia, but it was suddenly clear that genes represent the underlying mechanism for the causal connections. I have to get into the lab, I thought, and I spent the next ten years there. It was like the Cultural Revolution in China – all of us refined intellectuals had to go out into the field and learn some genetics."

Some interviews are easier than others. This is one of those interviews driven by the interviewee, because he has a story to tell, and my greatest contribution is to put the recorder on the table between us and nod understandingly at the right places.

"Strange," Weinberger says abruptly. "It is so strange to think that people just a couple of decades ago would not acknowledge that genes help determine the way we are. Everybody was fine with the fact that genes shape physiology and the body, but they repress the fact that the brain is also a part of the body – even traditional geneticists have a hard time with it. But, hell, we all look different, how in the world can we imagine that our differences only take a physical form?"

He shrugs his shoulders and screws up his eyes behind his rimless glasses.

"The reason we are a little afraid of individual genetic variation is because it is easy for value judgments to come into play. People will say that one variant is good, while another is bad. Good genes and bad genes – it sounds unpleasant, right?"

Perhaps it does, but at the same time it is impossible to get around the fact that there are human variations that are good to have in certain contexts and not good to have in others. When we look at different types of personalities, for example, is there is a natural repertoire?

"Yes, that much is obvious. It is also obvious that some variations in this repertoire are better in certain situations than others. And when we have spent a whole lot of time researching a particular gene…" He considers me with a certain expectation in his eye.

"You mean COMT?" I ask.

He nods like a smug schoolteacher.

THE GENE FOR catechol-*O*-methyl transferase, COMT, is and will remain Daniel Weinberger's signature object of study. It has become almost synonymous with his name and, again and again, his group publishes fresh findings about how variants of COMT work in the brain and influence the psyche. The gene codes for an enzyme that, among other things, breaks down the neurotransmitter dopamine in the frontal areas of the brain. These regions are central to cognition – everything involved in planning, reasoning, and conscious thinking.

The quantity of dopamine available to these areas of the brain is directly connected to the COMT enzyme. Greater enzyme activity leads to less dopamine production. The exact level of the enzyme's activity is determined by the specific sequence of COMT gene producing the protein. The version of the gene that has the amino acid valine at position number 158 in the chain imparts higher enzyme activity, in fact, four times higher than if the same position is filled by the amino acid methionine. This small difference has surprising repercussions, including, perhaps, to whether we live our lives as "warriors" or as "worriers."

"It's a question about the dose of dopamine in the cerebral cortex, and you can say that there are three settings," Weinberger explains. At one end, you find people with two copies of the valine variant, and they have the least dopamine available. This is reflected in physical effects that can be measured directly from brain scans. In general, these are people who function slightly worse cognitively – among other things, they do somewhat less well in certain memory tests. On the other hand, they are better at handling emotional stress. They have a higher pain threshold and go more directly after rewards in the form of activities that trigger increased dopamine in the brain. These are the warriors. They need more dopamine to feel alive. Weinberger clarifies: "The sorts of people who, in war, are up for attacking the enemy's machine gun nest and who look forward to the next stimulating battle."

At the other end of the spectrum are people with two copies of the methionine variant. These "met-met" individuals are cognitively more precise and, correspondingly, better at various memory tasks, but they handle emotional stress poorly. These are the world's worriers.

"I myself have two copies of the met variant," I note, because, of course, I checked this particular SNP in my Promethease report. "And I *do* worry. All my life, in fact…"

"Yes, well," answers Weinberger with a polite lack of interest. Personal twaddle is not his business. As a scientist, he must take a larger perspective – evolution and the human condition itself. He stresses that it is not as though one COMT variant is the right and righteous one, the other an evil, debilitating mutant. No, when it comes to behavioral genetics, we're talking about trade-offs between one area – emotional reactions – and another – cognitive tuning.

"You can show that, in evolution, there has been a genetic balance between effects that provide cognitive advantages and effects that provide emotional advantages. For every variant, there will be certain environments or niches where it is best suited. In our historical past, warriors were probably much better at hunting mammoths, while worriers stayed home in the caves and discovered fire."

Stories from old days and far climes are always pleasant, but I know there are meta-analyses that conclude that the different COMT variants provide no effect. It's a pet peeve of Weinberger's. His face puckers in response to my needling.

"*Meta-analyses*," he says as if he just bit into a lemon. "What's wrong with them is that they are mixing apples and oranges! Meta-analyses of COMT just smack together all the studies where they've done cognitive tests, even where the cognitive tests are not at all the same. And the other problem is that the analyses are strongly colored by one bad study that just happens to have a large number of participants. They don't take the quality of the various studies into consideration."

Weinberger asks whether I have heard about the new meta-analysis that, apparently, refutes Avshalom Caspi and Terrie Moffitt's study of SERT and depression. "Of course," I say with a clear conscience.

"This is a classic example. A meta-analysis that was ruined by one British study that did not find the effect Caspi had observed. They had over eight thousand participants, but the data were superficial, based on a single telephone interview. You can't compare work like that with a study like Caspi's, which followed a group over many years and where the researchers interviewed the participants repeatedly and face to face."

He bursts out again in his vigorous laugh.

"Know what? It's like what one of my geneticist colleagues says: if you check the Nobel Prize winners for the last hundred years, none of them worked with meta-analyses."

With that, I've run smack into the war between epidemiologists and molecular researchers, a long-standing enmity that Kenneth Kendler had mentioned to me in passing. One school thinks in terms of populations and percentages – in short, statistics – and they need data from groups, the bigger the groups, the better. The other side looks to the individual, trying to understand processes at the microscopic level that can be expressed in the differences observed across populations. For the molecular researchers, the point is to think small, to design well-planned studies that test a scientific hypothesis.

"If we are to understand anything about genes and behavior, we can't start from enormous population groups," Weinberger says dismissively. "Here's a good analogy you can put in your book."

I say thanks, but he doesn't seem – or choose – to hear me.

"Do you know why the big association studies have a hard time finding genes that have to do with behavior? Because they treat a particular example of behavior that covers many things as if it were a clearly delimited phenomenon. To think that 'anxiety' is a single phenomenon, biologically speaking, is … silly. It's like studying the cause of car accidents in the United States and just defining it as accidents with cars. The only thing you know is that the car is destroyed. You collect all the information you can – alcohol in the driver's blood, the driver's age, whether there was a woman next to him, how many years he had driven, the state of the tires, how old the car is, and so on."

Weinberger finishes the last of his soda and throws the can nonchalantly across the room, hitting it right into the trashcan.

"You conduct measurements all over the US. But because you are an epidemiologist and not a biologist, you look at a car accident as one thing and overlook the fact that there are many types of car accident. In Florida, where the average age of the driver is sixty-five, it's about elderly motorists; in Seattle, it's bad weather; down in the South, it's about the driver's alcohol level; and up in the Northeast, accidents are caused by your girlfriend sitting next to you, screaming at you and chewing you out. In the northwest, poor tires make the weather effect worse, while the same bad tires protect you in the Southwest, where the roads are hot and sticky. A risk factor in one place may be a protective factor in another place. And what do you end up with when you control in this way for five hundred different factors?"

I shrug my shoulders.

"That the primary cause of automobile accidents in the US is: *a driver's license*. Driver's licenses are the only common factor you can find in all those people, but this does not in itself have any predictive force. This is what the gene studies are coming up with for behavior and mental traits. The same for diabetes. Here, they always say that they have so much success, but they really don't; they have explained four percent of individual variance with their genetic risk factors. Four percent!"

Weinberger has just rehabilitated his colleague Caspi's controversial study of depression. But I would like to know what other findings they have for a connection between genes and behavioral traits. Are we talking about a lone swallow here?

"We have robust effects for brain systems. And brain systems are the foundation for behavior. We know from many studies that, under many conditions, SERT has an effect on how sensitive your amygdala is. It is a foundation for emotional engagement; in order to feel threatened or to feel anxiety, you have to have this activation of the amygdala. There are stimuli – sounds, strange faces, and that sort of thing – that must be there in order to stimulate your brain to produce some particular behavior. Genes are biological toolboxes that work on the configuration of your nervous system. They influence molecules and cells and thereby the whole structure of the brain, and the structure of your brain and its synapses determine how your environment is experienced and feels for you."

DURING THIS LITTLE sortie, Weinberger has acquired some color in his cheeks and looks ten years younger. His voice has become warm and well-oiled. And I know why. We have reached the point where he can explain how you can move beyond these despised genetic association studies.

"We need to focus on how *the brain* functions," he says triumphantly.

It cannot be denied that fascinating findings come out of looking at brain function. In a forthcoming study Weinberger's group examined the brains of one hundred normal volunteers to see if the COMT gene is linked to any distinguishing characteristics in the brain. They discovered that those individuals who carry two copies of the met variant exhibit a greater density of synapses between nerve cells in the anterior part of the brain than do others. Weinberger believes these extra synapses make a person more capable of concentrating – and may explain better results in various memory tasks. But there is a price: a psyche with *rigor mortis*. "They have a harder time shifting focus quickly and presumably also have a tendency to brood more on the ideas they have – including sad and negative thoughts," he explains.

Another conspicuous example of how the brain's function can be altered considerably by small genetic idiosyncrasies comes from research at the University of California, Los Angeles, led by psychologists Naomi Eisenberger and Matthew Lieberman. The researchers, who happen to be married, are investigating the neurology of social relations, and they have a hypothesis for why people with two copies of the less efficient MAOA variant may be more aggressive: the travails of this world hit them harder.

After identifying the MAOA variant for each of the volunteers in their study, Eisenberg and Lieberman put them into a brain scanner and subjected them to a computer game designed to make the player feel like an outsider, the equivalent of social rejection. The researchers then measured the activity in an area of the brain, the dorsal anterior cingulate cortex, which is central to assessing social situations involving psychological pain. In those who were endowed with the "aggressive" MAOA variant, the dorsal anterior cingulated cortex lit up like a firecracker when the player was rejected. Aggression was not, as had been assumed, the result of a lack of self-control. These were not insensitive brutes. They were super-sensitive, and their aggression rather an expression of psychological thin-skinnedness.

"MAOA is a really interesting gene,"Weinberger remarks, leaning back in his chair. "The first studies concentrated on aggressive behavior, which is just a reaction. But we gradually uncovered a picture that variants of MAOA more generally play a part in adjusting our social sensitivity." He sits up with a start. "We mustn't forget BDNF."

Hearing the words *brain-derived neurotrophic factor* is like encountering an old acquaintance. My PhD project looked at how the BDNF protein works in the brain. Of rats, that is.

"Sure, but it's the same thing," says Weinberger with a wave of his hand.

Whether in rats or people, BDNF is a *growth factor*, a protein that tells brain cells to divide or grow or create new synapses. Scientists have ascertained that BDNF exists in two versions and, just as with COMT, the difference comes down to one little amino acid, namely the one at position 66 in the protein chain, which can be either a valine or a methionine. When the BDNF gene codes for a methionine, the BDNF protein is pumped out of the producing cells at a slower rate, which means there is less growth factor available to the brain tissue. That *must* have significance. Yet scientists are still trying to discover precisely what that significance is.

One avenue of attack is to place subjects into situations that require different cognitive capacities. For example, researchers at the University of California, Irvine, looked at how driving ability corresponded to BDNF variants. A group of volunteers, all of whom had undergone genetic tests, were put into a driving simulator. Those with a BDNF gene with the methionine variant not only made the most mistakes, they also displayed a poorer ability to learn from those mistakes. Other studies claim that the variant is linked to negative effects in certain aspects of memory, not least what is called *episodic memory*, in which you recall autobiographical events. Weinberger's group has found indications that people with this same variant perform more poorly in tests in which they have to recognize words – "subtle effects, but measurable," as he terms them.

BDNF also appears to interact with SERT. In fact, in 2008 Weinberger and his colleagues published a report in the journal *Molecular Psychiatry* showing that a slow-release methionine variant of BDNF protects against some of the effects that are known to go with the short variant of SERT. If you are born with two short SERT variants, you typically have a harder time suppressing any negative emotional signals that are pumped out from the amygdala. These negative emotions loom large and, presumably, cause you to slip more easily into a depression. But if, in addition to your short SERT variants, you are lucky enough also to carry a defect in your BDNF genes, this effect is somewhat dampened. In other words, a variant that increases a certain risk in one genetic background can work protectively in another.

"E-p-i-s-t-a-s-i-s," says Weinberger, so everyone can follow along. "Interplay, you know, mutual control. We're sitting here talking about individual genes, right? And we are because the research typically looks at one gene at a time. But, ultimately, it's all about the interplay. I know Ken Kendler doesn't like this, but that's because he doesn't understand it. He keeps on claiming that data from population studies do not support epistasis, but it's not true."

Indeed, this may be the case with MAOA and testosterone. In one study, a team of American and Swedish researchers, led by Rikard Sjöberg at NIH, compared ninety-five Finnish alcoholics who had been convicted of violent crimes with forty-five law-abiding non-alcoholics. All the participants were psychologically assessed, completed a questionnaire designed to reveal their level of aggression, and had their testosterone levels and MAOA genes tested. The results indicated that those individuals with high testosterone and at least one MAOA variant with low activity had a greater tendency for antisocial and violent conduct. For those without this MAOA variant, testosterone levels did not seem to influence aggression levels.

Furthermore, a surprising study from the Caspi and Moffitt lab focused on attention deficit hyperactivity disorder. Though the ailment is known to be partially heritable, it can also express itself very differently in different people. Some hyperactive children not only have problems with attention but also with grossly antisocial behavior – they fight, they steal, they set fire to things. Caspi wanted to see whether there was a connection between such an antisocial element and the variants of the COMT gene. There was. The young people who had an ADHD diagnosis and also showed "early and consistent antisocial behavior" had, in the vast majority of instances, two copies of the "warrior" valine variant.

"It is probable," said Caspi when the study came out, "that the role COMT plays in relation to antisocial behavior touches on an interaction with other genes or behavioral factors in ADHD."

What happens when we can discover these sorts of connections, and more besides? Soon, when every child has a complete genetic map made at birth, these connections won't be the stuff of somewhat obscure scientific journal articles but everyone's everyday life. Would having this knowledge be good for us as individuals, or for society? If you knew from your earliest moments that both your copies of the COMT gene belonged to the met variant, making you predisposed to worrying and rumination, would it change the course of your existence?

Daniel Weinberger looks down at the table and squirms.

"Ah," he says at last. "You can't predict anything for the individual, because these are small buttons that are finely adjusted; the effect is not great from one single gene."

Of course, that is correct. You ought to be able to look at your genes and say, "Oh, perhaps I should do X out of consideration for Y," but how many of us can do that when confronted with our family history of cancer or some other disease? Honestly, I think he's avoiding the question. Because what scientist wouldn't want to be able to claim some benefit from his or her research, despite all the possible harms?

"Look at psychiatrists, you know, good, old-fashioned psychiatrists," he scoffs. "They used to ask people endless questions about their mothers, as if we could make good predictions on that basis. So, the profession really went off the rails by giving advice and making predictions on the basis of information that is completely subjective. At least, genes are objective. But, honestly, I don't know whether it helps you to know that you have a genetic disposition to be more fearful. That is something you already know about yourself from your own life."

"You mean it can become a self-fulfilling prophecy?"

"I just don't know whether you're served well by that information," says Weinberger, smoothing the bags under his eyes. "I believe, at any rate, that it is wrong to treat human behavioral variation as disease."

ON MY WAY out of the vast parking cellar, one of the guards lifts his hand from his weapon, and I can't help but wave back at him enthusiastically. I'm buoyant from my visit to Daniel Weinberger. Not least because his final remark runs so well with one of my own hobby horses.

It has long irritated me that our relationship to behavior and psyche is so deeply influenced by the concept that you can divide humanity into the sick and the well. This is a medical way of thinking. In a biological sense, you talk about *variation*, as Weinberger emphasizes. There are no healthy and unhealthy genes; variation is the modeling wax that evolution uses to create diverse forms. Mutations and their random alterations of the genome are a guarantee that evolution and adaptation occur. And variations that are not adaptive and impractical in one context may do quite well in another.

Asperger's syndrome is a superb example from the lived world of human experience. *Aspies*, as many call themselves, have a hard time automatically and intuitively understanding the emotions of others. They do not express their own feelings in the regular and accepted ways. They may seem awkward, volatile, strange. People with Asperger's syndrome are today classified as falling on the autistic spectrum; they display some of the typical autistic patterns of behavior and character traits, but in a high-functioning version. Those with Asperger's are poorly adapted to the furiously paced network economy, which seems to insist on constant, effortless social interaction and readiness for change.

On the other hand, other behaviors can accompany Asperger's syndrome. These include a tendency to nurture powerful personal interests, a drive to acquire an immensely detailed knowledge about them, and the ability to maintain an extraordinary mental focus. This aspect of being and character makes some Aspies brilliant at tasks that require contemplation and sustained concentration, especially if they are in an environment that provides the space for both and which ratchets down the need for social finesse. In this sense, the behavioral variation we call Asperger's syndrome is in no way "diseased."

Research in behavioral genetics has made it clear that genes do not exist in a vacuum but develop more or less advantageously depending on our circumstances. Such thinking is sure to change the way we see ourselves and one another. For instance, we often hear concerns about how more and more behaviors are labeled "sick." And if you look at the professional manuals on psychiatric illnesses, it's true that every new revision contains more and more specialized diagnoses. You could call it the clinician's diagnostic itch. Behavioral genetics stands in contrast to that trend: it shows us that the concept of "normal" must be expanded, not constricted, and that biological variation at the genetic level provides a broad palette of behavior and psychology. By demonstrating that genes do not *code* for behavior but *act* by adjusting and shaping our complex and dynamic nervous system, the pioneers who are combining genetics with brain research are uncovering how external influences and environmental factors create different outcomes for the brain – and thus for the person.

"The debate about nature versus nurture is over," the American psychologist Eric Turkheimer has briskly declared. "The bottom line is that everything is hereditary, and this is a conclusion that has surprised all camps in the debate."[26]

The logical consequence of this outlook should be not to focus narrowly on our genes. Rather, we need to use our genetic knowledge as a springboard for a greater understanding of ourselves, about how we inherit a certain quality, whether a disease or a behavior, by subjecting our personal genetics and our physical selves to a specific environment. By choosing to follow a particular diet or take a particular form of exercise, you may prevent an unfortunate combination of variants in your genome from expressing itself in the form of heart disease or diabetes. In the future, by partaking in special relationships or special mental or social activities, you may be able to dampen the effect of a gene that cranks your nervous system into an undesirable gear.

SO, HERE WE'VE finally reached that question of how much we'll gain when we fully grasp our genetic ballast. Daniel Weinberger isn't so sure it's a good idea to delve too deeply into our many mutations and alterations, and seems worried we'll get stuck on the idea that genes are fate. Of course, he's right that we are far from a sure recipe for how we should respond to the news that we carry a particular gene variant.

Yet, after my first taste of SNPs, my curiosity gets the better of my worrier reserve. I've filtered my raw data from deCODEme through the Promethease program and identified my COMT variants, but I want to know more. What about the other genes that we are beginning to realize have behavioral or psychological significance? Will I be able to understand myself and my idiosyncrasies and tendencies better if I look deeper?

I'm at least going to make the attempt. At home in Copenhagen, I've promised to make myself available to a battery of researchers who are trying to show how genes and childhood experiences interact to form personality itself.

6

Personality is a four-letter word

*We continue to shape our personality all our life. If we knew
ourselves perfectly, we should die.*

ALBERT CAMUS

"WE'RE TESTING FOR twelve genes. So, we need to fill that
many test tubes. And a couple extra as controls."

The young doctor plops a row of test tubes on the table and goes
out to get an appropriate needle to stick into my arm. We know each
other already from a previous occasion. She doesn't have the perky
ponytail this time, but she was the person who interrogated me
about my immediate family members – and their relationship to
alcohol and psychological illness – for a research project on personality, depression, and genes. Now, we're reunited in a basement
beneath Copenhagen University Hospital, to take care of the
genetic side of things. There is a major renovation going on, so we're
huddled around a low coffee table that's been incongruously placed
in the hallway, amidst the chaos of construction.

"You've got some really nice veins," she says, acknowledging the
slightly swollen blue vessels running down my left arm. They virtually offer themselves to the needle. Nevertheless, she somehow

manages to work it so that only a tiny trickle of blood is lured into the needle. She tries to improve the flow by nudging the needle back and forth inside the vein. It doesn't help – in fact, it hurts.

"Let's try the other arm," she says, quickly puncturing my right arm. The result is the same. "Seven years of training," I think to myself, "and what do they really learn in medical school?" I bite my tongue. In the end, I asked for this; I volunteered to be a guinea pig. The deal with the researchers at the Center for Integrated Molecular Brain Imaging is straightforward: they get a ream of completed questionnaires and copious amounts of blood from me, and I am allowed to discuss the results of my personal genetic tests with the center's director, Gitte Moos Knudsen, in return. I was able to extract this special favor because she and I share a fascination: we're "insanely interested in what creates individual differences in behavior and personality," as she puts it.

But what exactly is personality?

Here, you have to leave the molecular behind and move into psychology, where personality research traditionally resides. As opposed to molecules and neurotransmitters, which are made of atoms and chemicals, personality appears to be odd and ethereal. We have a clear intuition of what it is, but it is hard to find a crisp, sharp formulation. You might describe one of your friends as helpful, patient, and a little introverted, or yourself as bright, generous, and sociable. But squeezing the phenomenon of personality into a definition is a little like shoveling sand with a sieve. What in the world are you measuring?

And there is the burning question of where this indefinable quality comes from. Is personality readymade or is it created and shaped by life experience? If it is the latter, what shapes it – the actions of a parent, for good or for ill? The myriad random acts that accumulate in a lifetime?

If you ask parents, you will typically hear them say that their child's personality was apparent from birth. "Little Laura was hot-tempered from day one" or "Harry had the most serene temperament right from the start." In other words, we think we know very quickly what sort of child we're dealing with and, in a sense, that means we can just sit back and watch that personality unfold over the years. There is not much to be done about it one way or the other.

Yet, ever since Freud described childhood as the key to all understanding, psychology has scoured early experiences and influences for answers to pretty much everything – just as we continue to do in our everyday lives. We happily refer to events, moods, and feelings from decades ago to try to explain our reactions in the here and now. We look for patterns in our thoughts and behaviors that derive from our childhood among our parents and peers. A bad relationship with a long-dead mother gets the blame for his introverted and sour personality in middle age. Growing up in the countryside with a huge horde of siblings is the reason she is vociferous and domineering at home and at work.

But how reasonable is this?

This was a subject I regularly debated with my father – though neither of us had anything but vague assumptions to support our arguments. In his later years, he developed a bad conscience about practices he considered, after the bulk of his child-rearing was complete, to be parental sins. All his adult daughter's problems, he believed, must have their roots in her upbringing. Was it, perhaps, his overwrought expectations that meant I had chronic issues with dissatisfaction and unfulfilled ambitions? I wasn't so sure. My father retorted that unquestionably, it was too early for me to start learning the alphabet when I was still being toilet-trained as a two-year-old. Well, I couldn't remember any of that. But in a faded snapshot, I'm sitting on the potty with lettered wooden blocks strewn before me spelling the word C-A-T. I don't look unhappy. Sure, it may have been pushing things a bit to insist that I should be able to read before nursery school, or that our morning bike ride to the school was spent calling out chess moves, back and forth – playing "blindfold chess."

My mother thought this sort of thing was madness, bordering on child abuse. Yet, though there were quarrels over when I should be put to bed, during daylight hours I was my father's daughter, his accomplice in my upbringing. It probably helped that reading was accompanied by rewards: every time a whole sentence of *Dick and Jane* was read aloud, I got a nice, plump raisin; when an entire page was complete, I'd get a piece of candy. And praise, of course. Since then, like the proverbial Pavlovian dog, I have associated reading with the ingestion of sugary substances. I practically salivate at the sight of a book.

Still, I maintain that none of this harmed me. Not as such. I don't go around with a feeling that anyone forced me to do anything I didn't want to do. To my guilt-ridden father, I used to point out what happened when I finally toddled off to nursery school, *Dick and Jane* tucked under my arm: I discovered that my fellow three-year-olds couldn't read as much as their own name. So I promptly renounced all my training. Neither raisins nor caramels made any difference. In the end, it was far more important to be like the other children than to do what Daddy wanted.

These days, in my adult incarnation, I gladly argue that, on the whole, childhood should not be burdened with culpability for whatever goes wrong in life. Or right, for that matter. As children, we are stuck with the parents we have and the upbringing they give us, and both can be a trial. But at some point, we take over the controls for ourselves. Thereafter, life is pretty much our own responsibility.

My father wanted terribly to accept this, but he never bought it entirely, perhaps because he did not himself feel that he had broken free of his parents or his childhood, to which he traced much of his own personality, especially the more difficult aspects of it.

Who is right? How does this work? What can you say about the source of your personality and its biological mechanisms?

THESE ARE SOME of the central issues being plumbed in the field of personality research, which is experiencing a renaissance. In good part, that resurrection is driven by the fact that scientists have finally come up with a good model for personality and can use genetics to explore it. No longer are researchers stuck fighting over ideology-tinged theories, from classical Freudianism, to psychodynamics, to social psychology. With the genome available for investigation, personality research is joining the ranks of true science.

The number of the beast is the *five factor model*. As the name implies, the model identifies five general personality traits, or personality dimensions: openness, conscientiousness, extroversion, agreeableness, and neuroticism. These 'Big Five' dimensions describe a range of behavioral tendencies in everyone's character. Neuroticism is the tendency to lose yourself in negative impulses, while extroversion is the tendency to seek company and be confident. Openness is the tendency to engage in new and different views and experiences, while conscientiousness is the tendency to plan, be dutiful, and exercise restraint rather than spontaneity. Agreeableness speaks for itself.

Each of the five dimensions is composed of six facets, which could better be described as specific traits statistically associated with the general dimension. Warmth, gregariousness, and excitement-seeking are facets of extroversion. Self-discipline and order belong to conscientiousness. Hostility, depression, and anxiety are tied to neuroticism.

Despite the neat boxes and categories, this does not build up into a watertight, idiomatic definition of personality. Instead, many psychologists purport that we are dealing with a set of basic *dispositions* that outline a general pattern for how we think, feel, and act and how we react to the world around us. You can also put it more abstrusely. For example, Daniel Nettle, a psychologist at the University of Newcastle, in the UK, describes personality traits as "stable individual differences in the reactivity of mental mechanisms designed to respond to particular classes of situation."

You have to imagine that all humans are equipped with the same basic feelings and reactions. How easily these are engaged in each of us and how powerfully they unfold varies greatly. People with a neurotic personality react especially easily to the register of negative emotions, which they also experience very powerfully, while a personality characterized by extroversion experiences a much stronger response to positive emotions.

Contrary to a multitude of other personality tests, the five factor model does not derive from a chromium-plated theory of personality; it is data-driven, a crystallization of measurements and observations from laboratory experiments and real life. Oddly enough, like behavioral genetics, it can trace its history back to good old Francis Galton. Darwin's talented and enterprising cousin desperately wanted to measure human "character," as he called it, and he had the excellent idea that features of our personality must necessarily appear in language: in the way we speak about each other. Therefore, Galton went to the dictionaries and collected somewhere in the range of a thousand English words that described facets and nuances of character or personality. By classifying synonyms, he boiled the heap down into far fewer unique personality traits, which in 1884, he published in an article entitled "The Measurement of Character."

The modern reinvention of personality research grew out of the same lexicographical method. In 1936, the British team of Gordon Allport and H.S. Odbert burrowed through contemporary English dictionaries and came up with a list of 4500 adjectives that they believed fully described all possible human personality traits.[3] But things only got moving with measurements of actual people in the 1950s and 1960s, when researchers began to let subjects characterize themselves as well as others. To do so, subjects would often link a number score to the many descriptive words.

With this breakthrough, some order could be glimpsed in the apparent cacophony of the psyche. It became clear that certain aspects of personality go together. Using *factor analysis*, a tool of mathematics that can determine how a huge quantity of data can be grouped together and made coherent, researchers could reduce the many ingredients down to an essential few.

The first person to make use of this statistical method was Raymond Cattell, cofounder of the University of Illinois' Institute for Personality and Ability Testing, who discovered evidence for sixteen different, independent factors in data from a personality test he had developed. That work laid a foundation for later investigations of personality data. Then, in 1961, Ernest Tupes and Raymond Cristal of the US Air Force Personnel Laboratory applied factor analysis to data from eight different major studies, revealing the five factors for the first time. Their identification of the factors was confirmed just two years later by a colleague, Warren Norman of the University of Michigan. Throughout the rest of the decade, the dominant idea among researchers was that personality traits could be distinguished and measured; that there was something tangible about a human being's internal makeup which could be used to predict behavior.

But then the tumultuous 1970s got in the way. A powerful ideological wind was blowing, and it shook personality research. Now, the social psychologists had their turn, and they believed an entity as reactionary as a "stable personality" did not exist. Leading figures in the field, such as the American psychologist Walter Mischel, argued that who we are is dependent on circumstance, and personality traits are behaviors and attitudes we adapt in given situations. Some went so far as to say that personality is nothing but a construct, something we invent about others and mentally impose on our view of them – to create at least an illusion that the world has permanence.

After the social psychology revolt, the early pioneers of personality research – Allport and Odbert, Tupes and Cristal – were largely forgotten, but science reported once again to the field. This time, the prominent psychologist Lewis Goldberg, dictionary in hand and factor analysis in back pocket, blazed his way to the five dimensions of personality and coined the "Big Five" designation. During the 1980s, the model won broad acceptance among psychologists, and people finally had a common language with which they could discuss personality.

A consensus emerged that you could describe an individual's personality in a coordinate system involving five axes, one for each factor. You are not simply extroverted, agreeable, or neurotic – you contain some traits from every single dimension, but are more or less characterized by one or another. Your score placed within these five dimensions, so to speak, could be considered to be your overall psychological chart, akin to a series of your physical measurements. Chest, waist, hips, for example, or height, weight, and body fat.

The American researchers Paul Costa Jr. and Robert McCrae distilled and decoded the five factor model in their whopper of a book, *Personality in Adulthood.* Published in 1990, it is still the closest thing there is to a Bible of personality. The duo also constructed a personality test that has gradually become the gold standard. The complete *NEO Personality Inventory-Revised,* or NEO-PI-R, introduced in 1992, consists of 240 questions – or, rather, statements – that are assessed on a scale from 1 to 5, in relation to how well they fit the subject. These are simple, psychologically elegant, statements, such as:

> *I have excellent ideas.*
> *I am not interested in abstractions.*
> *I am the life of the party.*
> *I don't like to call attention to myself.*
> *I insult people.*
> *I am easily disturbed.*
> *I am always prepared.*
> *I have a soft heart.*
> *I change my mood a lot.*

I may not have the softest heart around (I'd give it a 3), or be the life of the party (though I never understood why my excellent jokes often seem to chase people away). But do I perhaps insult people? Well, no more than to demand a 4, I should think.

THE FIVE FACTOR model has its detractors. Some are piqued by its very simplicity – they argue that just five dimensions are too rough a measure to say anything meaningful. Others question whether the five factors are truly independent; some research indicates they may not be. If they are not independent, the model might eventually be reduced to a four factor model or a three factor model, leaving personality psychologists with a very limited palette of "genuine" traits with which to explain the multiplicity of human personality.

At the moment, however, the five factors rule. Several studies have shown the factors to be surprisingly stable, meaning that when measured against the Big Five, a person's personality does not fluctuate up and down, like the stock market or average rainfall curves. For instance, over six years the dynamic duo of Costa and McCrae followed a group of people who sat down at regular intervals for a tête-à-tête with a NEO-PI-R test. The researchers found that an individual's score varied very little over time and, in fact, the difference between two measurements taken six years apart was no greater than that between two measurements taken six weeks apart. Later analyses and meta-analyses have found the same thing – but with the addition that it looks as though there are overarching, long-term patterns in the five factors of personality. Over the course of a lifetime, levels of neuroticism, extroversion, and openness go down, while levels of agreeableness and conscientiousness increase a little. An observation, in other words, that fits very well with the old maxim that we mellow with age as "the rough edges wear off."

Another indication that the model captures reality is the ease with which it lines up with distinct personality disorders. This is supported by an interesting study conducted by the American researchers Lisa Saulsman and Andrew Page. They asked individuals who had been diagnosed with one of ten categories of clinical personality disorders to take a Big Five personality test. When they analyzed the results, they found that they could accurately predict how the patient would score on the test solely from his or her diagnosis.

So far, so good. But the ultimate test of a theoretical model is whether it is able to predict practical conduct and life *as it is lived*. Remarkably, it looks as though the Big Five can say something, statistically, about the course of human life.

Extroversion is a good example. People who score high on this dimension, who are typically outgoing and sociable, and who have easy access to positive feelings, also appear more likely to be employed in extroverted jobs. To a higher degree than average, they work as salespeople, and hold positions of staff responsibility. Extroverts also consistently have more sexual partners than the average person, just as they have more marriages over the course of their lives.

In turn, we can say something about how a marriage works if we get information about spouses' scores on the personality test. Statistically, divorce is far more likely if either the man or the woman has a high neuroticism score. A man with a high conscientiousness score is an indicator that the marriage will last, though interestingly, the woman's conscientiousness score seems to make no difference to the relationship. You have to wonder if this pattern reflects the fact that most of these studies were done years ago, with long-married couples of a generation in which the husbands were still the providers.

Certainly, your conscientiousness level is a strong indicator of how you will perform in school – no matter what your sex. You might believe that the most important ingredient is raw intelligence but, as psychologist Maureen Conard of the University of Connecticut reported in 2006: "Aptitude Is Not Enough." Conard reviewed the personality scores and the classroom marks of

American university students and found their level of conscientiousness was a better predictor of future grade average than the entrance examinations, known as aptitude tests, that the students had taken before being accepted.

Conscientiousness also says something about how people do after their happy-go-lucky student days. A 1991 meta-analysis by the American researchers Murray Barrick and Michael Mount, who considered personality tests of more than twenty-three thousand subjects following a great many different professions and positions, established a clear and straightforward connection between a person's conscientiousness and his or her job performance. No matter the performance criteria, the more conscientious among us did better on average than the less conscientious – regardless of other qualifications.

Even though many of these personality analyses have involved Americans, it appears the five factor model is quite culturally robust. Its concepts and questionnaires have been translated into numerous languages and the findings hold up across the globe. It is even the case that, across cultures, you can see the same patterns of sex differences. Everywhere, men on average score higher in extroversion and conscientiousness than do women, while women on average score higher in neuroticism and agreeableness. Indeed, women's scores on agreeableness run so high that the average man will score lower than seventy percent of women.

"REMEMBER TO ANSWER quickly without thinking too much about it," was the instruction from the research head as I prepared to haul myself through the full version of the NEO-PI-R. I followed orders, making my way through statement after statement, knowing that after I had finished the test would be analyzed and scrutinized by the psychologist Henrik Skovdahl Hansen, who is an expert in just this sort of personality interpretation.

As the director of the Danish publishing house Psychological Publishers, he has not only translated and taught generations of psychologists about the NEO-PI-R, he also wrote his PhD thesis on the use of personality tests. I have only spoken briefly with Hansen on the phone, but when I meet him at his offices, he matches my expectations for a business psychologist quite nicely: no shaggy hair or penchant for Indian scarves. He is wearing a discreet, dark navy suit and sports short-cropped hair and narrow steel-framed spectacles. On the other hand, his handshake is more hesitant than I had anticipated.

"*You're* Lone?" he asks, looking at me in a slightly noncommittal way before he shows me into his brightly lit office. A secretary efficiently places glasses of coffee and some mixed chocolates before us. I immediately grab a bite-size Snickers bar. Hansen touches neither food nor drink. Instead, he folds his hands in front of him on the table and begins the session by professing his faith in the five factor model and its anchoring in a "scientific paradigm."

With these five measurable factors and their facets, scientists have finally constructed a model of personality that is so well founded that it reaches beyond pure psychology. They can study the connection between personality traits, biochemical processes in the brain, and the genetics that lie beneath them all. "Of course, biology is not exactly my field, but it is the future," predicts Hansen.

Before we concentrate on my test, there is one thing in particular he wants to impart. "The five main traits or dimensions themselves, on which the model is based, are not particularly interesting, because in the very nature of things most people lie somewhere in the middle. It is only a minor portion of the population who are extremely neurotic or very extroverted, but from the more specific aspects of each dimension, you can really say a lot about a person."

I don't know whether this is good or bad in my particular case. So I smile to be on the safe side. Hansen then stresses that, while personality traits are quite stable statistically, this applies across the population; anything can happen as far as the individual is concerned. The professionals who work with the test, he explains, know from sheer experience that when people take these tests again and again there are some who change radically. A person may even encounter things in life that push him or her far off the normal distribution.

"Let me be more precise," he says. "A person who is very introverted does not suddenly become very extroverted, but sometimes you see significant changes."

"Does this typically occur after major life-altering events?" I ask, thinking of accidents, illnesses, death. "Or is it rather because they themselves have decided to do something?"

"You quite often see that psychotherapy makes for changes in personality. On the other hand, you typically see that major life events do not influence personality. It is rather your personality that colors the way such an event affects you. Grief, for example. I believe that personality largely determines how grief over losing other people plays out in the individual."

I sense this could lead us too far astray, so I inquire whether we should look at my test. Without further ado, Hansen pulls it out and points to a green schematic diagram cluttered with boxes and numbers, brutally dominated by a zig-zagging graph that cuts through them.

"This graph is made from your scores on the 240 questions and a comparison of this score to the norm. A score of ten or twenty is in itself meaningless; you have to know where you are situated within the different dimensions and facets in relation to a comparable group of people to get any knowledge from it. In your case, we have chosen an overall norm for businesswomen. That is, hundreds of Danish women who are employed at all levels from top to bottom."

Hansen points to the boxes where the five dimensions are broken down into their six facets. "Here, you get higher resolution. You can typically see that people score in the middle of the general extroversion dimension, but this middling value may reflect the mean of a high score in the facets of gregariousness and warmth and a low score, for example, on excitement-seeking."

I understand but want to know what sort of person appears in the curves on my graph. What sort of person *I* am.

"Yes, yes," Hansen replies, looking as though he is considering his answer carefully. I guide him to a sharply declining line, which falls far off the norm indicated by a transverse green band running across the middle of the page.

"It looks as if I'm scoring pretty low on agreeableness."

"Yes," Hansen says gently. "You are. In fact, you *can't* score any lower."

This fact hangs in the air between us for a moment.

Hansen now admits that he had not been looking forward to our meeting, once he had analyzed my data. "But let's look at it. You're part of an experiment that touches on depression, and let me tell you that your test does not provide any basis whatsoever to talk about a depressive personality. You see, it is crucial to distinguish between a condition and a personality trait. Depression is a condition that you have a disposition for if you score high on the trait of neuroticism ..."

I feel it is necessary to break in. It may be that I am no more neurotic than the average Danish working woman, but I must emphasize that, after all, I have had three episodes of clinical depression. Clinical depression *requiring treatment*.

"Yes, you told me," Hansen responds, remarkably level. "But from your relatively average score on neuroticism, one wouldn't say that you are especially inclined toward depression."

I inspect the graph, and it is true enough that my neuroticism is just lying there, resting on the upper boundary of the normal area. That is unexpected. Then, Hansen pulls the paper toward him and points to a facet of neuroticism that is actually labeled "depression".

"Here, too, you are somewhere in the middle. That is, all the people you are being compared with on the average have answered exactly the same as you with respect to having a dark view of things. If you are to interpret it, you would say that, when *you* react with sad-ness and depression, there is usually an external reason for it. On the other hand, if you score high on depression, this sadness would have a life of its own."

To me, this sounds exactly as if Hansen is describing the differ-ence between what psychiatry used to call melancholy, which comes from within, and reactive depression, which is provoked by external circumstances or influences. At the same time, I also sense that he is prodding my self-image. Because, quite honestly, I have always considered myself an incurable melancholic. Someone who suffers because she is naturally predetermined to suffer but, by virtue of some inner strength, nevertheless bears it more or less sto-ically. But is this a lie, a self-deception? Am I really just a weakling who cannot tolerate the small jolts of life?

"Well," says Hansen, getting me back on track, "if you look at all the facets of the neuroticism dimension, it appears that you don't worry more than other people. On the other hand ..."

He clears his throat. Twice.

"... your threshold for when you become irritated ... You get there much faster than other people, and you can get really angry and have difficulty controlling your anger."

Sorry, but give me a break. Doesn't the man realize how stupid other people can be?

"When it comes to traits we know are involved in depression, it is the *anger component* that is most pronounced in you," he says, examining me through his narrow glasses.

"You see this prominently in the test?"

"I do. And I wouldn't be surprised if most people you know said the same thing about you."

They probably would; people are so quick to judge. But I sense we've become stuck on a single issue and try to move the conversa-tion to other subjects. There must be something encouraging to latch onto.

"I actually think it's a little strange that my score on extroversion is down in the low area, below the average. Can that really be the case?"

In my reading of the literature on personality, I've fallen a little in love with the extroverts, who sound like they experience a lot more pleasure, all around. They get more sex, and get it with many more different partners. They also rake in a higher salary.

"I can't understand it. Because one of the things I really think is fun is speaking in front of large audiences. That must be extroverted, right?"

Hansen smiles indulgently.

"You still haven't quite grasped the dynamic involved in a personality. I'll say it again — it's no good to stare blindly at individual dimensions. Another element that is interesting in precisely this connection is your very low score on social anxiety. You say that it doesn't bother you to get up in front of an assembly?"

"No, I actually like it."

"That is your social robustness coming through, and that is quite excellent. But on the minus side, it can make you quite insensitive to how you affect others."

Suddenly, a lot of things make sense.

"But your social robustness has to be compared to another score, in your openness dimension, on what we call depth of emotions. It is high, above average, as you see, and it tells me that your openness to your own feelings and those of others and your ability to empathize is actually high. So, here, I see a conflict," Hansen says, sounding as if he enjoys it. "How does it actually work, with you being easy to irritate and, maybe, even to piss off royally, and not to catch the signals from others even though you actually have the antennae to be able to catch them?"

What can I tell him? I could break down and confess that I frequently get into quarrels with people; that I have a habit of saying something I know is going to upset them but, for one reason or another, I can't stop myself. I could disclose my tendency to fire off rude e-mails, which I later regret but have difficulty pulling back.

"When I look at my upbringing," I say, instead, "it had to do with never worrying about what others might think or say about you. My father taught me never to doubt that I was right and to just go out into the world and do what I thought was best." In fact, that part about doing what I thought best was very important.

"That corresponds quite well to the profile you show in the test: what I hear you saying I can see on paper," Hansen comments. "Your agreeableness couldn't be lower, and your compliance is rooting around down at the bottom – that is, there is a desire to debate things and some competitiveness. Then, there is your low score on altruism and sympathy. This does not mean you can't feel empathy, but that you have opinions and stand by them and that you don't shy away from making unpopular decisions," he says, making me wonder whether I should have considered a career in management.

"It is no accident that you are interested in science. That is typical for people with a low sympathy score. If you had been high up, you would probably have done something in the humanities." He hesitates a moment. "In fact, you have some quite masculine traits for a woman."

Funny how many people – unbidden – seem to want to enlighten me about that.

"Quite honestly, you must really get at loggerheads with people."

All the time, as it happens.

"This is something that nourishes your personality; you need struggle to keep yourself going, it drives you," analyzes Hansen. "But if you want to talk about *development*, here is where you'd grab hold, because you can also see that you can take things hard and be quite sad. Your sensitivity to stress, which is a vulnerability factor, is high, and you also have a depth of emotions that makes you vulnerable. You are not cold but …"

"Combative?"

"A fighting machine, I would say. If you want to say something positive, these things keep you from getting into situations where you can really step on other people's toes and not notice it."

"You mean that, without sensitivity to stress and depth of emotions, I would be a really unpleasant person?"

"You could interpret it that way."

We sit silently for a bit, and I allow myself another candy from the bowl. Then, I focus on what Hansen just noted about development, which is, after all, my true mission in meeting with him. I want to find out how much our genetics defines personality, and how much of it we can develop as we choose.

"As I said, personality is normally quite stable. And that is why the five factor model says, at heart, that you need to make peace with your personality. Accept it and try to shift it around a little," he begins. "I don't think you can make a frontal attack on your personality, but you can refurbish it a bit and, maybe, become better at what you think you're bad at or, at least, cover it up."

He offers me an analogy, comparing personality to a sports team. If you have a team with a weak player in one position, you can't simply go out and acquire the world's most expensive player to replace that person – even a top-level professional club has to make the budget work. Instead, to accommodate your weak spot, you switch around your players or try new tactics. The same idea applies to developing personality.

When it's time for me to leave, Hansen gives me the decorative test graph as a souvenir and supplies me with a computer-generated report of my test answers, as well as an accompanying personality analysis. There are diagrams galore, but it also contains some good advice. I take special notice of the conclusion on the last page and carefully highlight it in yellow to give to my boyfriend back home:

> You are more vulnerable and susceptible to stress and strain than most people. It is important that you have supportive people around you to lean on.

VULNERABLE BUT DISAGREEABLE. This is unquestionably a bad combination – whether you consider it from the inside or the outside. But where does it come from?

According to science, I can put a large part of the blame on my parents and the genes they passed on to me. When you investigate the evidence, there is a surprising amount of genetics in our personality. Generally speaking, as measured by the Big Five test, personality is around fifty percent heritable, and a major meta-analysis of previous twin studies indicates that the individual dimensions vary in heritability from forty-two percent for agreeableness to fifty-seven percent for openness. That means the impact of genes and environment is more or less equal.

Yet, the environment that matters may not be what you expect. "Behavioural-genetic research provides the best available evidence for the importance of environmental influences, but it shows that the environment works in a surprising way," writes the psychiatrist Robert Plomin in a thorough review of the topic. You'd expect that your childhood home and the way your parents raised you would have a huge influence, but the environment that really matters seems to be anything *other* than what your parents create.

Initially, this conclusion faced an uphill battle. In fact, when Plomin and his colleague Denise Daniels first published this conclusion in their 1987 article "Why Are Children in the Same Family So Different from One Another?," it hit psychological circles with shock and awe. By analyzing previously conducted twin studies and adoption studies, Plomin and Daniels showed that neither personality nor psychological development is shaped by the environment siblings share when they are growing up. At least, that is, if their upbringing is not marked by outright abuse or neglect. This point was most clearly illustrated in research with children who, as toddlers, had been transplanted into adoptive families and raised with children unrelated to them by biology. If upbringing strongly influences personality, you would expect that children raised together would resemble each other more in personality than two random people off the street. But they don't. The environment that influences personality, argued Plomin and Daniels, is what is called *non-shared environment* – the environment you don't share with your siblings.

Nailing down what counts as non-shared environment has proved difficult. One of the better proposals comes from an American researcher, Judith Rich Harris, who in her groundbreaking and controversial 1998 book *The Nurture Assumption* argues that as children we are colored by our *peer group*. The large component of personality that Mom and Dad are unable to influence is shaped by our same-age friends, with whom we try with all our might to fit in.

Harris's idea gained support from work by the Newcastle University psychologist Daniel Nettle. In *Personality: What Makes You the Way You Are*, published in 2007, he offers a twist on the peer group approach, claiming that the environment's influence on our personality has to do with the world's reaction to us. Our appearance and our intelligence affect how we are received by others, and the reaction of others is reflected back and incorporated into our personality. This ping-ponging helps regulate where, on a sliding scale, the five factors or personality dimensions are placed for an individual. When tall men, for example, score lower on agreeableness and higher on extroversion than comparable men of average height, it is because they do not need to please anyone, Nettle points out. Most people, as a starting point, are positively inclined toward tall men.

All this is not to say that parents mean nothing – they can, indeed, cause their children lasting harm from the way they choose to run their family. Parental influence can easily shape the way you are within the family throughout your life. But as Nettle puts it, "The point is that this does not generalize to the adult personality with which the offspring addresses the rest of the world."

How am I to understand my own social robustness – or disagreeableness, if you will? Right off the bat, you might say, echoing Henrik Skovdahl Hansen, it corresponds precisely to my upbringing, which therefore must be its cause. But it could also come from my biological heritage. Perhaps I simply inherited some of the genetic elements that also made my father indifferent to the opinions of those around him. Or, to make it even more complicated, could it have to do with the fact that my genetic disposition fits particularly well with the way I was brought up?

"IT WILL JUST be five minutes, but come in and take your coat off."

Gitte Moos Knudsen is trying to get three major funding applications sent off before the deadline in ten minutes' time. Still, she looks remarkably tranquil sitting at her computer, and the spacious Copenhagen University Hospital office, with abstract art on the walls and Arne Jacobsen furniture swathed in light grey leather is virtually immaculate. The only signs that the young professor has had a rough day are her slightly reddened eyes.

"Oh, it used to be worse when they wanted a dozen copies of each application," she says, beaming with vigor and equanimity. I happen to know she is the type who bikes the eight miles to and from work – even when it is so cold that the gears freeze up. So when she asks me, after a little, whether I would like some warm milk in my coffee, I feel like a poor specimen of the human race.

What has she seen, I think, as I take the cup. As research director of the Center for Integrated Molecular Brain Imaging, Moos Knudsen possesses a lot of data on me: a brain scan, a battery of tests on my cognitive abilities and social understanding, twenty-four hours of measurements of the stress hormone cortisol, and questionnaires on everything from sleep rhythms to personal problems. I've already learned about my skills in cognition – memory, verbal capacity, that sort of thing – and there is nothing to worry about there. I've also learned, quite surprisingly, that my social understanding lands at the high end of the spectrum. This afternoon, I've come to get the results of my genetic analyses. It's nerve-wracking.

To bolster myself, I decide to share a quotation that I picked up from the British psychologist Wendy Johnson: "Understanding genetic mechanisms for personality traits is one of the biggest mysteries facing the behavioral sciences."

Moos Knudsen nods and pulls her chair up to the round conference table.

"We're interested in behavior and personality, because they are also risk factors for psychological illness," she starts. "We are beginning to see traits and facets of personality that suggest a propensity for illness. One of the best described links is between a high neuroticism score and a propensity to develop depression and anxiety disorders. But we also know that people who score low on agreeableness have an increased risk for cardiovascular disease." As an afterthought she adds "They also die earlier."

A little dry in the throat, I mention my own personality test and my own miserably low score on agreeableness.

"I see," says Moos Knudsen. She seems undisturbed. "To put it briefly, there is interplay between genes and environmental factors that together determine who we are as people. Our ambition is to understand the intermediate stations found on the way from gene and environment at one end to behavior and personality at the other."

One of the intermediate stations she and her team are investigating is the structure of the brain, particularly its intricate chemical communication systems of neurotransmitters, which are exchanged in quick impulses between cells; receptors, which capture and pass on the signals; and hormones, which circulate and provide their own subtle messages. A sloshing ocean of overlapping mechanisms is involved, but the team is especially interested in the actions of serotonin.

"It is so marvelously versatile," she says of the neurotransmitter.

Indubitably. This little molecule is not only significant in the regulation of basic needs such as sleep and appetite but also exercises its effects in more complex matters, including humor, aggression, sexual behavior, and the way we react to punishment and loss. Many of the things that are said to color personality. And when I look down at the list of the genes Moos Knudsen has been so kind as to test in me, they are all pieces in the great serotonin system: genes coding for proteins that transport serotonin in tissue, or for a number of different receptors that, each in their own way, convey signals via the same neurotransmitter.

Take the receptor with the prosaic name of 5-HT_{1A}. It is a big, intricately folded protein, which sits snugly on the surface of cells in almost every structure of the brain and is deeply involved in a broad repertoire of our cognitive abilities. It is also the most widespread of all the serotonin receptors. Scientists know from experiments that, among other things, the 5-HT_{1A} receptor plays a role in both long-term memory and attention, and genetic studies have shown that it is found in many versions, each of which shifts, ever so slightly, the receptor's function.

"The most studied variation is the SNP we call rs6295," Moos Knudsen informs me, as she riffles through her papers. In this variation either a C or a G nucleobase appears in the 5-HTR1A gene in a certain position. The specific nucleobase affects how the receptor reacts to the serotonin swimming around in the brain. So far, scientists have determined that a receptor with the G variant signals less efficiently.

"People have had ideas that this precise gene variant has significance for various psychiatric illnesses, but the studies have not really been able to conclude anything," Moos Knudsen mumbles. Then, she finally finds what she is looking for. "Here it is! Your gene test. I can tell you that you have two copies of the C variant."

This news is almost cheering. I'd stumbled onto a trailblazing approach to genetic research suggesting that people with two C variants think less rigidly and with less conformity than people with either two G variants or one copy of each. Curiously, the results also highlighted that how gene variants influence behavior depends on the cultural context in which they've grown up. The woman who designed the unusual study is the Korean-born Heejung Kim, a psychologist at the University of California, Santa Barbara, and her take on the genetics of personality has prompted the trendy magazine *Seed* to call her a genuine "revolutionary mind."

Her main work is centered on cognitive flexibility in relation to 5-HT1A. Psychologists know that this receptor plays a central role in how well we adjust our mode of thinking in response to our circumstances. Kim's assumption was that there would be a difference in performance depending on which variant a person carried. The G variant, with its less efficient signal, would provide less flexibility and thus a poorer ability to alter cognitive style, compared to the C variant. If that is the case, those people in every culture who think most traditionally or most in sync with mainstream culture would be likely to be those with the G variant. Kim confirmed this by comparing Koreans with Americans.

We know from anthropological and psychological studies that there are characteristic differences in thinking and self-image between Eastern and Western cultures. Whereas Eastern cultures are consistently more community-oriented, Western cultures are more individual-oriented. For example, when Korean, Chinese, and Japanese people are asked to describe a picture, they emphasize, for the most part, the background and the totality; Westerners confronted with the same task typically focus on the foreground and the details. Researchers also talk about holistic thinking, prevalent in Eastern cultures, versus analytic thinking, prevalent in Western ones.

Kim presented a group of Koreans and a group of Americans of various ethnicities with a standardized questionnaire that determines the degree to which a person's thinking is collectivist or individualist. Then, she tested each individual's 5-HT1A gene. When the results were compared, Kim's hypothesis held water. In both cultures, the individuals with two G variants were the most "culturally typical": those Koreans who thought in the most collectivist ways and those Americans who thought in the most individualistic

ways. By contrast, people with two C variants were the least culturally typical. Snugly in the middle were those individuals carrying
one of each type of variant.

"Fascinating," remarks Gitte Moos Knudsen when I share the
study's findings with her. "Here we see how different cultural patterns can very directly influence how the same gene gets expressed
in behavior, right? On the whole, I think the interplay between culture and genes will be a huge theme in future research. It's enormously exciting."

Personally, I'm warming to the idea of my efficient 5-HT1A
receptors and I'm just about to entertain Moos Knudsen with how
well I think the result fits my own level of mental flexibility.
However, she is already scanning down her list. She pauses for an
instant before announcing: "Then, there is the gene for BDNF"– the
growth factor that makes brain cells divide and form new connections and is therefore a part of attuning how the brain responds to its
surroundings.

You test for two variants. One codes for a protein with the
amino acid valine in a certain position, whereas the other codes for
methionine, and methionine has the unfortunate effect of reducing
the amount of BDNF available for brain cells to bathe in. The
methionine variant is rare, as the vast majority of people have two
copies of valine. Moos Knudsen reveals that I carry one of each.

"I just knew it!" It bursts out of me, and she looks somewhat surprised. From my bag, I produce an article that has just been published in *Psychoneuroimmunology*. In it, a research group from the
Hebrew University in Jerusalem describes how women who happen to be equipped with at least one methionine variant of the
BDNF gene appear to be worse at handling stress than those with
two valine variants.

The research team asked just under a hundred subjects to act
out a job interview to the best of their abilities – but with cameras
rolling and spotlights glaring in their faces. Following this ordeal,
the subjects had to solve a bunch of mathematical problems. During

the process, each had their production of the hormone cortisol (produced in response to stress) measured. Among the men, the production of cortisol – and, thus, the individual's stress level – rose most for those with two valine variants of the BDNF gene. In contrast, among the women, those who had the greatest stress response were those with one copy of each variant.

"Interesting that there is a gender difference," says Moos Knudsen, flipping through the article. "I wonder what the mechanism is?"

The Israeli team had no credible proposal to answer her question, so we move to another sex difference that is applicable to one of my genes: the gene that codes for the enzyme monoamine oxidase, or MAOA. As you will remember, MAOA breaks down the neurotransmitter serotonin, and variants of the MAOA gene that lower the production of the enzyme have been linked to a person's higher sensitivity to social pain. Rejection, social isolation, and similar experiences simply activate more powerful negative emotions in those with less efficient MAOA.

"Everyone used to call it the 'warrior' variant, because they were focusing on aggression," Moos Knudsen observes. "But it looks like men and women express increased sensitivity differently. Men react in a more extroverted way with aggressive behavior, while women tend to turn the reaction inward."

Depression, thy name is woman, I think instinctively before asking nicely for my own results. They don't look so good – I carry two copies of the less efficient variant. Both Mom and Dad delivered.

AFTER I'VE HAD a microscopic chance to absorb that news, we turn to good old SERT. This gene for the serotonin transporter is probably the most studied gene in psychiatry.

"The extent to which the short variant produces a particular vulnerability or not is being debated now," Gitte Moos Knudsen begins. "I believe the current thinking is this: you can have two copies of the short variant and still go through life in an excellent mood without a single depression. But it is a vulnerability variant in the sense that, combined with unfortunate life circumstances, it is not good to have."

I wait while she studies her papers.

"You do have two copies of the short variant," she says, undisturbed when I throw my pen on the table.

Of course. I *knew* this. I sit there, pondering the genetic fiasco: two short variants, the worst conceivable combination. So, my damned recurring depressions don't just come from nowhere, I grumble, thinking back to my meeting with Henrik Skovdahl Hansen and his neuroticism index. As if it explains anything, I tell the story about how my family boasts at least three suicides, when I count both sides. Successful suicides, that is. While I'm grumbling on and on about being cheated of the more "robust" gene variants, I notice that Moos Knudsen is smiling wryly.

"You have to ask yourself what the sensitive variants are good for," she says then.

"Good for?" I don't really follow her point.

"Take the short version of SERT. If it were really only a disadvantage, why has it survived many millions of years of evolutionary development, and why is it so frequent? Almost one in five Caucasians has two copies, like you. Have you thought what advantages there might be with that variant?"

Thought *and* thought, that's overstating it a bit. But asked directly, I offer up a vague idea about how fragility can make you warier than more thick-skinned people in situations where things might possibly go wrong. It is conceivable, I suggest, that those of us who have our antennae out to sense life's brutality are also better at reflecting on our impressions of the world. Moos Knudsen nods slowly and asks whether my personality test doesn't show a high score on the openness dimension. I confirm that it does and am swiftly handed an article.

In it, Moos Knudsen and her colleagues propose that openness is the dimension linked to cognitive flexibility and a risk of depression. They suggest that short SERT variants pull an individual in the direction of increased sensitivity, but also seem to provide for a less rigid psyche. In fact, the scientists have seen indications of an interesting interplay between the SERT gene and age. Even though research subjects with and without the short variant could point to the same high score on openness, those who possess the short variant appeared more likely to preserve their openness over time, while others slack off on this personality dimension.

"Another thing is that people with greater openness are better at coping with psychological suffering, just as they have a higher survival rate when it comes to physical illnesses," she adds. I will comfort myself with this finding the next time life takes a dark turn. And when I'm hit by cardiovascular disease because of my unusually low score on agreeableness.

"But there is more. Over the past few years, there has been an increased focus on a personality type defined by the American psychiatrist Elaine Aron: people who are characterized by *sensory processing sensitivity,* or highly sensitive people. As may be obvious, they engage in more reflection and with more depth, which can be an advantage in some circumstances. It is not proven but it is a theory that is worth investigating."

According to Aron, the highly sensitive are found all over the personality universe of the five factor model. Some are very extroverted, while others are deeply neurotic. Yet, all can be characterized by the way they react to the environment around them. They take a longer time to process information but have more sensibility to detail and subtle distinctions. They have a higher sensitivity to sense impressions, such as loud noises and repulsive smells, and a lower tolerance for stress and unpleasant situations. Aron and her fellow researchers believe that these hypersensitive types could make up as much as a fifth of the population.

Moos Knudsen explains that she and others are starting to

investigate the early life experiences of highly sensitive individuals, including their relationships with their parents, to get a better understanding of how the personality type works. "Some findings actually indicate that, with a good and healthy upbringing, these highly sensitive souls do not just avoid problems but generally do *better* than anyone else," she says. This pricks my ears.

"They turn out to be amazing artists or people who really develop themselves in other ways."

She looks at me.

"Do you think you might like to fill out a couple of question-naires? They concern sensitivity and the parent–child relationship. You could be one of our first research subjects on the project."

Sure – let's finally get it all out on the table.

It turns out the study involves about fifty questions, to be answered on a scale from 1 to 7, and it doesn't take me long. As soon I finish, the completed forms are passed to Cecilie Löe Licht, a young PhD who is responsible for the study. She disappears into an adjoining room and promises to analyze my answers at once.

I fill the waiting by inviting Moos Knudsen to consider what all our poking around with the genetics of personality will lead to over time. Right now, behavioral genetics still only boasts a small set of circus horses that are, time and again, led by the nose around the ring to perform a variety of tricks. Scientists' knowledge of this handful of genes is based on relatively few studies, and it has a purely statistical foundation. But will she and her colleagues one day reach a point where they will be able to use a genetic profile to advise indi-viduals on how to conduct their lives? Will they be able to say, for example, that a child should avoid one sort of environment and seek out another?

At first, Moos Knudsen is hesitant. She wriggles in her chair a bit.

"A few years ago, I was probably more optimistic about that idea. There are presumably many more variants involved than the ones we study today, and the combination of variants is undoubtedly important. To tease this out would require enormous studies."

But they are also about to get underway, I fire back. In Europe, for example, the IMAGEN project gathers researchers scattered across the continent who are following two thousand teenagers over four years. The teens are regularly subjected to brain scans, psychological tests, and questionnaires about their lives and their doings. They are also being tested for a large array of genes. The researchers hope to find some patterns that predict the appearance of psychological and behavioral problems and to head those problems off through a treatment based in biology. "We conduct this study in order to better understand the teenage mind," says the consortium's majestic mission statement.

Moos Knudsen folds her hands on the table. "At a minimum, you can put together a genetic profile that will say something about how a person is equipped for various circumstances," she acknowledges. "In psychiatry, people are working at full throttle on prevention. And if you can find something that acts as a risk indicator for mental conditions as well as for physical disease, it would be good. Really good."

She becomes very quiet and still for a moment, then abruptly leans forward over the table and presses both hands to her temples.

"I'm constantly surprised, but when I'm out in public lecturing on genetics and personality, there are always people who get angry. 'I don't want to be my parents,' they say. As soon as people hear the word 'heredity' or 'genetic,' it immediately gets transformed into 'unchangeable' somewhere in their mind, and that is not the message at all. A personality is a product of genes and environment, and even though we cannot, of course, freely choose who we are, we still have a certain latitude. Throughout our lives."

There is a discreet cough from the door, where Cecilie Licht is waving her analysis of my questionnaire. With a muffled voice, she assures me that I can confidently count myself among the highly sensitive.

IN THE DAYS after my visit to the grey villa nestled among the concrete behemoths of Copenhagen University Hospital, I digest the information I brought home. At first glance, it looks like a catalogue of unfortunate gene variants. If only I weren't obliged to take them all.

I can count the variants on my fingers. There is the COMT gene, where I have a double dose of the "worrier" variant, which inclines my brain to handle emotional strain poorly, and my BDNF variants, which turn up my reaction to stress. A bad combination. On top of that, there are two copies of the less efficient MAOA variant, which disposes a person to aggressive and impulsive behavior – or depression, in the case of women. Finally, I am saddled with two copies of the short SERT variant, which is a notorious guarantor of psychological vulnerability and a tendency to depression. Add it all up and it sounds like the recipe for a walking psychological abrasion. Or as my boyfriend puts it: "It's remarkable that you haven't ended up in a closed ward or an early retirement."

You have to constantly remind yourself that genetic studies contain a healthy degree of uncertainty and only provide a statistical digest of reality. It is not a given that such findings apply to me particularly or anyone else who goes off and gets a gene test. Nevertheless, it seems pretty plain that I am a pitiful loser in the genetic lottery, one of those people who, due to the malevolence of fate, got an overdose of sensitivity and risk packed into her biology.

And yet, not. Or, at least, not necessarily. For as Gitte Moos Knudsen indicated, scientists are developing a new way of interpreting the phenomenon of genetic sensitivity.

In the earlier days of behavioral genetics, there was a blinkered focus on vulnerability, as researchers put a glaring spotlight on those poor wretches who happen to carry an unfortunate genetic burden and then run into a horrible childhood, and thus are inevitably subject to psychological illness throughout their lives. But as Jay Belsky, a child psychologist at London's Birkbeck College, points out, this does not capture the full scene.

Belsky argues we should be thinking in terms not of vulnerability but of *susceptibility* – or, more precisely, of *plasticity*. When it comes to genes such as SERT and MAOA, variants that heretofore have been characterized as making a person "vulnerable" or "at risk," need to be considered as making that person (and his or her nervous system) more sensitive and flexible in responding to relevant information from the surrounding environment. This sensitivity, in turn, makes a person more susceptible not just to negative influences but also, and to the same degree, to positive ones – a genetic tendency that is highly plastic rather than one-dimensional.

The framework of vulnerability and risk has until now been so absolute that otherwise excellent researchers have missed plasticity when it is found in their own observations. Belsky and his colleagues combed painstakingly through data from well-known studies, including Avshalom Caspi and Terrie Moffitt's Dunedin research, and uncovered something interesting: while it is true enough that the "vulnerable" variants of SERT and MAOA, when combined with child neglect and even violence, increase the risk that a person will end up with depression and behavioral problems, the reverse is also true; as soon as you're talking about a normal childhood, the carrier of these variants will have *fewer* depressions and *fewer* behavioral problems, compared to those with more robust variants. In other words, the sensitive do better than the robust. They get more out of the absence of stress than the robust. They thrive. In almost poetic terms, the psychologist Bruce Ellis and pediatrician Thomas Boyce have described the theory as the difference between "dandelions" and "orchids." The former are the robust among us, the salt of the earth, those people that carry the species through even in adverse conditions. The latter are those who wither when the gardening is deficient but who blossom wildly and beautifully with the right care.

So, the question remains: what provides this much-coveted normal or even above-average childhood? As the researchers at Copenhagen University Hospital noted, a close bond between parent and child may have a decisive effect. When you probe the scientific evidence, you can see some of the first indications that this is the case.

For instance, in 2009 a joint team from Columbia University in New York and the University of Pittsburgh discovered that the quality of parental care directly curbs the effect of the vulnerable – that is, less efficient – variant of the MAOA gene. The study makes for captivating reading. The researchers asked 159 adult women, all of whom were diagnosed with depression or bipolar disorder, to grade their parents on the quality of their parental care. In addition, the women were asked to provide information about early traumatic experiences, such as divorce, death in the family, physical violence, or sexual abuse. They were also evaluated psychiatrically and scored for aggressive and impulsive tendencies. Finally, they were tested for the make-up of their MAOA gene.

In the study, the researchers showed that those women equipped with vulnerable MAOA variants were the most sensitive to the stressful experiences of childhood. A childhood marked by trauma could make these women more aggressive and impulsive in adulthood than women in the genetically robust group. But they were also the only ones who showed any response to a high degree of parental care; for some of the women, their tendency to develop aggressive and impulsive behavior seemed to be attenuated by positive parenting. Conversely, the genetically robust women, all of whom had highly efficient MAOA variants, gained no special benefit from what they themselves described as good parental care. For them, stressful events led to a relatively high score on aggressive and impulsive behavior regardless of how their parents treated them.

As I pore over the literature, I'm inspired to call Cecilie Licht to find out more about the project on the highly sensitive personalities among whom I am now enrolled. The young researcher explains that she wants to study how a suite of gene variants influence our personality dimensions and how they work together with the early parent–child relationship. I am reminded of Robert Plomin and his much-touted model, which claims that our childhood home environment plays a negligible role in shaping our personality. But Licht refers to some promising new ideas, including a theory suggested by Jay Belsky that this only applies to the genetically robust among us; the genetically susceptible *can* be influenced by the environments our parents create.

How does a scientist go about measuring parental environment, especially decades later? The relationship between parent and offspring is complex, and it seems impossible to boil it down to a simple, fifty-question form. Yet, Licht and her colleagues in this research focus on two general aspects of personality that, they posit, are crucial: the extent to which a child experiences care or rejection; and whether the child feels autonomous, or restricted and overprotected. As Licht herself formulates it: "You can imagine a scale in which cold control lies at one end, and warm freedom at the other."

I can't help but think of Dad. How we banded together over reading and raisins and how we could always talk about everything like adults, even when I was just a kid. Certainly, there was a nice collection of stressful events in my childhood, but there was also an enduring relationship with my father. I could talk about unconditional love and care, but "warm freedom" is actually a very precise way of putting it.

"How can I lay down the law for you, when you are a far more reasonable person than I am," he said on one occasion. I was nine then, and there *were* no prohibitions. No fixed bedtimes, no special times when I had to be home, no forbidden television shows. That sort of thing my father believed his sensible daughter could easily manage. And when my self-management went awry, it was remedied with an ironic finesse.

Like the time two years later, when I was caught in a minor shoplifting incident with a friend and co-conspirator, and we were held by the shop owner, who wanted our parents called in. Naturally, it was my father we phoned and, of course, he came and retrieved us and talked the furious lady down to earth. When my friend dissolved into tears, convinced that her volatile mother would tear her limb from limb, my father drove her home and talked her mother down to earth, too. In the car along the way, he practiced his own form of pedagogy. "Girls, I hope you've learned how stupid it is to steal candy bars and funny erasers. If you'd swiped a nice Bang & Olufsen television or a stereo, you could respect that, but this is just embarrassing."

My snuffling friend in the back seat just gawked, and once again it was confirmed that my father was the greatest adult in the world. Now, I can only speculate whether he actually saved me from the worst effects of the sensitive genes I inherited from both my parents. Maybe susceptible variants are not the genetic bullets they appear but, in my particular context, a beneficial tool. At one point, Licht says what I am thinking myself: "It's entirely possible that you did better with your sensitive variants than you would have done with the more robust ones."

It is possible, yes, but no one can say so with full confidence. Behavioral genetics can provide some intriguing indicators, but we are still dealing with research that scratches the surface of personality. If we still know so little, what does this information mean for us, here and now?

"Personal genomics has a long way to go before it will be a significant tool for self-discovery," says the psychologist Steven Pinker. And, he is right, in the sense that the molecular answers will not surprise us with revelations of traits we didn't already know about.

I have lived with myself every single day for over forty years and am not blind to the flaws and tender spots of my character. Stress, for instance. Before I subjected myself to personality testing and genetic analyses, I understood that the way I dealt with stress and strain was not something to be proud of. After more than ten years as a professional journalist, I remain chronically bad at deadlines. While my hardboiled colleagues take them with aplomb, I can freeze when a sudden time pressure arises and, in a few seconds, transform from a seemingly stable adult human being into a hysterical wreck.

So, what difference would it make to hear about some exotic gene variants that push a bit on the serotonin system? I'm not sure I agree with Pinker on this. Even now – in spite of all the uncertainty and lack of knowledge – it makes a difference to self-discovery; it provides a biological awareness that can very well affect your self-image.

I dig out an old *New York Times* clipping I had almost forgotten about but which has acquired new meaning. It is an account of two American sisters, Tichelle and La'Tanya, and how their common starting point led to drastically different destinies. The girls really drew the short straws in the great parent lottery. Their mother was an alcoholic, their father was out of the picture, and their stepfather, who was installed in the household, abused the girls sexually from a very young age. The girls, in other words, shared a horrendous childhood.

As a young adult, however, Tichelle seemed to thrive. She completed her education with good marks, and put together a life as a computer operator and as a parent who can be said to be a success, all things considered. Her big sister La'Tanya, on the other hand, struggled. Though she, too, was a mother and had a vocation, as a nursing assistant, from time to time she fell into a deep hole of depression. She experienced anxiety attacks and had difficulty holding onto a job and, more generally, dealing with the challenges of everyday life.

At one point, the two sisters, at the suggestion of an entrepreneurial reporter, had their genes tested. The hardy Tichelle had two copies of the long SERT variant, while the fragile La'Tanya carried one long and one short version. "I feel a little better that there is a reason, another reason, for my life being hard," La'Tanya said when she heard the results. "And I understand that what I'm able to do for myself and my kids, even with this, is good. It's good."

No behavioral geneticist would claim that this little variant is itself the decisive cause for the young woman's situation. But it is probably a piece in the puzzle, and La'Tanya uses this knowledge to forgive herself for the fact that she has to fight hard to, quite simply, feel better. You could say that she leans comfortably on a molecular crutch.

For me, the question is whether this sort of crutch, in some hands, might not help carry us forward. In other words, whether a person could use genetic insight as a lever with which to change the self. Statistically, we know that personality is quite stable but, as Henrik Skovdahl Hansen indicated, great changes can occur in the individual, if the person is engaged in the process. Then, the statistics may actually mean that either most people do not find it worth the effort to make the attempt, choosing the motto "I'm wonderful the way I am," or they do not try to change themselves thoroughly enough.

But the chances for being able to change are better the more we know about the preconditions of personality. And a good number of the preconditions are tucked into our genetic heritage. If I know about the existence of a number of gene variants that probably help pump up my stress reactions, I can use that information in various ways. I could, for instance, do as La'Tanya did and focus on dealing better with my own incapacity, to choose, with a clear conscience, to turn down the heat. Nothing wrong with that. But maybe I don't need to avoid stressful situations. Maybe, instead, I can think my way into responding better to them. When things are too hot, I can tell myself that it is not necessarily the world that is unmanageable, but my brain which, in part because of its genetic influences, is viewing the world that way.

For this, I might take a cue from a recent study by Dina Schardt and her colleagues, of the University of Bonn, who suggest "genetically predisposed neural processing may be counteracted by willful actions." According to their research, you can whip your hypersensitive emotional brain into submission by exerting some cognitive control. Inspired by older findings that the short, "sensitive" SERT gene variant makes the amygdala more jumpy when you confront something fearful, the team got hold of thirty-seven women who were genotyped for SERT and put them in an MRI scanner, where they were shown a set of standardized pictures designed either to be neutral or to evoke fear. Just as Daniel Weinberger and others have seen, the carriers of short SERT variants reacted much more strongly to the fear stimulus than did the others. But that was only until the next scanning session, when the women were asked not just to look at the pictures but to volitionally detach themselves from the sight – a trick they had been taught and practiced in advance. When confronted with the fear-inducing pictures, both genetically sensitive and genetically robust women were able to dampen the activation of their amygdala. In fact, the two groups became indistinguishable. The sensitive individuals more efficiently suppressed their feelings of fear, showing that even if you are predisposed to have a stronger immediate emotional reaction, you can regulate it through conscious cognitive effort, if you so choose.

So when the going gets a bit tough, I can remind myself that my genes do not directly affect behavior but the mechanisms of my nervous system. And though my genetic inheritance plays a part in setting the conditions in which my brain operates, it is an incredibly plastic organ, whose chemistry is constantly influenced by how it is used. By thinking the right way, I can develop a thick mental skin to surround and protect my hypersensitive physiology.

7

The interpreter of biologies

*DNA is just a tape carrying information, and a tape is
no good without a player.*
BRYAN TURNER

"WHAT DID YOU say this was for?"

My general practitioner looks suspiciously at the two thick plastic cylinders with built-in needles that I have brought with me, and the detailed instructions about how they are to be filled with blood, turned slowly ten times each, and then quickly sent off for analysis to a research laboratory at Lundbeck, a Danish pharmaceutical multinational, located on the outskirts of Copenhagen.

"They want to diagnose mental illness with a blood test?" the doctor asks as she routinely thrusts one of the needles into my vein. Yes, they do, but the method is currently in development. The idea is to move away from the psychiatrist's habit of diagnosing various ills through subjective assessments of more or less ethereal symptoms. You ask patients how they think things are going, relying on a checklist of foggy questions about emotional states, sleep patterns, and psychomotor functions to reach an overall evaluation. Depression, for example. Or social phobia. Or borderline personality disorders.

Instead, the clinicians at Lundbeck and elsewhere are pining to discover *biomarkers*, objective measures comparable to those doctors use when they register the quantity of sugar in the blood to diagnose diabetes, or apply ingenious electrodes to the body to see if a heart is out of whack. Biomarkers are the Holy Grail. So when I catch scent of the fact that Lundbeck is trying to test the activity of selected genes to diagnose conditions such as post-traumatic stress disorder, borderline personality disorder, and, especially, depression, I leap at the chance to become a research subject one more time.

"But you're not feeling depressed at the moment?" my doctor asks, gazing at me with a look that says, "you know the drill, better to tell me now."

I shake my head, because I really feel fine. None of the familiar symptoms – waking up in the morning filled with a loathing for life; experiencing the day as a wearying drudge between two periods of sleep; sensing that everyone else in the world is doing exceptionally well and that I'm the only hopeless loser. For the time being, my mood sits in the normal range, which is to say, tolerable. It's far from the point where the film usually breaks, and I politely ask to renew my prescription for 150 milligrams of antidepressants to restore my sanity.

Reassured, she wades into the specifics of the tests, which are unlike any of the genetic tests that I've taken in the past. In contrast to Gitte Moos Knudsen and her fellow brain researchers, the people at Lundbeck are not interested in the particular, immutable sequence of my genes, but in how my organism chooses to interpret them. To get this information, they isolate white blood cells and calculate their number of RNA molecules, which are transcribed from a group of selected genes and provide a measure for how much of the corresponding protein can be formed.

This isn't traditional genetics, it's *epigenetics*.

"THE AGE OF epigenetics has arrived," *Time* magazine solemnly proclaimed a few days before my visit to the doctor. In a similar spirit, one of my American acquaintances, who works in a lab at Harvard, characterizes the field as "hot shit." Or as the biologist Denise Barlow of the Austrian Academy of Sciences lyrically puts it, "Epigenetics is about all the strange and wonderful things that can't be explained by genetics itself."

Such language is understandable, because it is presumably in epigenetics that the almost mystical encounter between inheritance and environment will be realized. It has been easy enough over the years to identify a genetic predisposition and then imagine the environment entering the picture, creating interplay and a result: a phenotype. But what does that interplay consist of, and exactly when and where does it take place? This is where things become more difficult.

Epi – that elegant little Greek prefix hints that we are dealing with something "above" or "beyond" genetics. The concept was invented in 1942, by the British biologist, Conrad Waddington, to describe how an organism's experiences and circumstances might make its genetic material act differently. At that time, before the genetic code was even revealed, it was all ideas and theories. Today, the field of epigenetics has come to stand for the investigation of how genes are *expressed* – that is, how much or how little protein they are allowed to produce, at what time, and in what cells. These are changes in the function of genes that occur without mutations in the genetic sequence.

Until relatively recently, however, most scientists believed that epigenetics was not relevant for adults – it was a phenomenon largely restricted to embryos, where the genome was programmed in ways that would govern the rest of the organism's life. After all, the purpose of developing from a single cell to a complete organism is, practically speaking, so that all cell types possess the same genome, though each uses a greater or lesser portion of that material. Depending on which genes are allowed to be activated, each cell acquires an identity, with a corresponding function, both of which are defined by the set of proteins the cell's genes produce; rather like an orchestra in which all the musicians have the same score in front of them, but the violins, the kettledrum, the triangle, and all the rest, play a distinct part.

Take, for instance, a liver cell and a brain cell. The liver uses a battery of enzymes to break down the toxins you consume when you eat and drink, and which must be removed from the blood. There is no reason for the brawny liver cells to produce the full array of complicated receptors that facilitate communication among the nervous system's somewhat more refined cells. Thus, the genes responsible for such receptors are inactivated in the liver. Inside the cranium, however, the brain cells are free of the garbage collector function, so the genes that specify enzymes that break down alcohol or fat are relegated to eternal rest.

The body's epigenetic program organizes this division of labor. It does so by turning specific genes on and off, in essence by making a single gene accessible or inaccessible to the whole cellular apparatus, which is necessary for copying the DNA into peripatetic RNA for production of protein. One effective way of making a gene inaccessible is to put molecular obstacles in its way. In practice, this happens when extra chemical connections – methyl groups – are attached to selected bases in the DNA strand, preventing the transcription of the relevant gene. However, accessibility also has to do with how the DNA molecule is packed. The forty-six chromosomes of the human genome do not lie relaxed and stretched out in the cell (if they did, they would run to two meters long), but are tightly wound around particular proteins, reminiscent of hair around rollers. These proteins – *histones* – can be modified chemically in ways that make them more or less tight and the DNA strand more or less accessible.

It turns out there are a bunch of specialized enzymes that make such modifications either to DNA or to histones, and just as many whose job it is to remove those very modifications. It also turns out that these enzymes are found in cells throughout life. Therefore, it is not strange that scientists have recently discovered that epigenetic re-programming takes place from cradle to grave and, presumably, in all of the body's tissues.

As is often the case in genetics, a good place to hunt for evidence is in identical twins. Although twins are said to be born with the same genome, they are never entirely the same, either in the way they look or in the way they think. Some of this is due to mutations, but some is probably the result of epigenetic changes. In 2005, a group led by Mario Fraga at the Spanish National Cancer Research Centre tested this hypothesis in forty twins between the ages of three and seventy-four. The researchers looked for patterns in epigenetic modifications of blood, muscle tissue, and skin in order to compare each set of twins and twins at different ages. They found a progressive development toward more difference. As small children, the twins were pretty much indistinguishable, but the more time they spent in the centrifuge of life, the more significant their differences became, and those differences were nicely spread throughout the entire genome. Epigenetics thus explains why one twin gets more wrinkles and ages more quickly than the other. Why one is fatter than the other. Or why one gets arteriosclerosis or schizophrenia, while the other does not. Reviewing these data, it seems likely that these different outcomes come from different lifestyles, since in infancy the lives of twins are more uniform than they later become.

If you think about it from an evolutionary perspective, it is completely understandable that these sorts of ongoing adjustments can take place, the enthusiasts of epigenetics point out. Presumably, epigenetic programming is a tool we have developed because it provides better possibilities for surviving. It is an adaptive mechanism that can change the individual in accordance with the requirements of the changing environment. The ingenious off-and-on mechanisms might even be considered an individual's personal capacity for evolution.

Of course, the changes may take an unfortunate turn and become pathological. Many forms of cancer, for example, appear to be caused by epigenetic modifications that make cells divide uncontrollably, giving birth to malignant tumors. For this reason, cancer researchers number among the vanguard in epigenetics. Recently, they have been joined by scientists searching for the roots of psychiatric illnesses in the brain.

Consider the characteristics of diseases of the psyche. They typically involve a significant genetic disposition, yet, to be realized, require some contribution from environmental effects – but no one can put a finger on what exactly those effects are. Psychiatric illnesses and syndromes also have other significant common features: they are accompanied by long-term behavioral changes; often develop gradually; and, when treated, it takes a long time for the symptoms to abate. The need for chronic medication is typical; this isn't a case where a simple chemical imbalance can be quickly fixed, once and for all. Finally, some of the medications that seem to stabilize the mood of both depressive and manic patients actually affect processes such as DNA *methylation* – a central component of epigenetic programming. All signs point to epigenetics at work.

But can something as ephemeral as social experiences – indefinable influences from your upbringing and interactions with people – really increase or decrease the activity of your genes? Researchers have come across clues that these influences are real. What gave them the clue was a pack of unfit mother rats.

In 2004, Moshe Szyf, of McGill University in Montreal, observed some interesting behavior over several generations of rats in his lab. He noted that rat babies that were raised by uncaring mothers – which, in the case of rats, means mothers who licked their babies only rarely and groomed them poorly – developed changes in the way they reacted to stress. Szyf could see plainly that the mistreated rat babies grew up to display much more fear than the offspring of good mothers. When the fearful female rats in their turn became mothers, they also neglected their babies, while the daughters of good mothers grew up to be excellent and attentive parents. The behavior was not inherited, but a direct result of upbringing, because if you took the newborn babies and switched them around between good and bad mothers, the babies took after their adoptive mothers. This was proof of a direct environmental effect.

What the environment had done came to light when Szyf and his colleagues killed the adult rats, plucked out their brains, and studied them in detail. It turned out that a poor childhood left lasting traces in the area of the gene for the *glycocorticoid receptor* – an especially important player in both rats and people. This receptor is crucial for the regulation of the stress response, because it helps prevent the formation of stress hormones when we are under duress. Among the mistreated rats, Szyf discovered, the gene had been completely blocked off by appended methyl groups. Parental neglect appeared to turn off particular genes in the rat's brain cells, even when the rat had otherwise robust genes.

Ultimately, the Szyf team was interested in whether the same thing held true for people, since it has long been known that children who have been subject to neglect, abuse, or violence develop abnormally strong stress reactions as adults. These same children also have an increased risk of depression and suicide, among other things. Here was the chance of a lifetime to find a direct explanation for where sensitivity to psychological illnesses comes from. The only problem: getting hold of some brain tissue.

So Szyf went to the local brain bank – the Quebec Suicide Brain Bank – and asked whether they might have some tissue from individuals who had committed suicide and who had been abused as children. They did. To begin, Szyf and his team were given tissue from twelve people who had committed suicide, all of whom had been subjected to sexual abuse, regular violence, or gross neglect, which they compared to tissue from twelve people of the same age who had unluckily died in accidents. They spotted clear differences between the two groups in the area of the hippocampus. The gene for the glycocorticoid receptor had been turned off, via methylation, in the cells of the abused – just like in the rats. There was also much less RNA from the glycocorticoid gene, which indicated that the gene was not very active.

The question was whether the genetic change was directly associated with the earlier mistreatment or with the fact that the individuals had been depressed enough to kill themselves. Szyf requested brain tissue from another twelve suicide victims who had not experienced abuse but had been depressed. And it turned out that this group did not stand out from the control group of accidental deaths. The particular epigenetic modification of the glycocorticoid receptor in the hippocampus was akin to a signature for mistreatment, the sordid fingerprints of abuse in the victim's brain.

You can get a similar signature from depression – your mother's depression, that is. It is well known that children of depressed mothers themselves have an increased risk of recurring depression and, for a long time, psychiatrists attributed this to the learning of depressive behavior. But it looks as though this sensitivity can be traced earlier, to the embryo.

According to research by Tim Oberlander of the University of British Columbia, there is an epigenetic effect on a child's glycocorticoid receptor when a woman experiences depression during the third trimester. This is the same signature found in Szyf's suicide victims, but this time it was measured using blood cells from the umbilical cord. When Oberlander and his colleagues later tested the three-month-old children, he found they reacted to stress by producing significantly more cortisol than the offspring of non-depressed mothers. The researchers could even ascertain that the epigenetic effect was the same whether the mothers had been treated with antidepressants or not.

Epigenetics has emerged as one of the most hopeful areas of twenty-first-century genetics research, and this is particularly due to the fact that the field seems to provide some long-desired explanations for how our environment mediates our genes. But the hope also rests on something else: that all the epigenetic changes are, in principle, *reversible* – something can be done about them. This stands in glaring opposition to the mutations we are so used to hearing about, which cannot be changed. Could epigenetics mark the dawn of a better world? Is it possible we can intervene with drugs and turn genes back on, or hinder the activity of those running wild?

"WE HAVE HAD some technical problems, but now the assay is up and running again, and your test results have now arrived from the laboratory. If you are interested, please call me."

This laconic e-mail has arrived from Birgitte Søgaard, the head of Lundbeck's department for translational medicine. Her department is the one that is supposed to translate the results of research into something that can be used outside the sheltered laboratory environment. On real patients. I have not been in contact with Søgaard before – I just delivered my blood test to a lab technician, who sent it on to the research team in New Jersey – but I call at once. Of course, I'm interested.

"Hello," chirps Søgaard, who sounds like she's in an especially good mood. Without further ado, she informs me, rather too cheerily, that my test places me solidly in the group of depressed research subjects.

"What?" I'm a little confused by this. I didn't feel depressed when I offered up my vein, and I don't now. This does not bode well for their test, I should think.

"You don't have any symptoms?"

The usual dissatisfaction and occasional frustration with my immediate circumstances, but nothing abnormal.

"Okay. I want to stress that the test here is in its early development phase and that your result need not be a mistake. Not at all. But, perhaps, you should come out here and get a more detailed explanation."

I agree at once, but unfortunately I won't be able to meet with Søgaard herself, who is about to leave on a trip to the New Jersey offices. After some negotiation back and forth, her colleague Jennifer Larsen, who is a biologist involved in this major test project, makes herself available for that afternoon. All I need to do is show up.

Several hours later, I step across the threshold into the special domain of the pharmaceutical industry. This is top-flight research in business class. There is no trace of the eternal struggle against the poverty and cutbacks that permeate the halls of academia and give universities a telltale atmosphere of anxiety. Here, there is room and money enough.

Through the yellow buildings in Valby, you are sluiced into a gigantic glass lobby, which opens into another hall as high and wide as a train station. It is empty of people except for two receptionists, who are situated in a sort of screened-off playpen, where they look like discarded Lego figurines. Equipped with a guest pass, I traverse the reception hall to one of the luxurious chairs plunked into the corner designated as a waiting area, where a high-end Bang & Olufsen flat screen television supplies the news of the day. The floors are silky dark shale, punctuated by vases as tall as a man, positioned here and there to provide a feeling of security as well as decoration. Even a granite sculpture of a human brain does not seem nerdy but aesthetically elegant — and expensive.

"May we speak English?" I hear from behind me.

Jennifer Larsen is Canadian, and even though she has lived here for five years and can easily speak Danish, she doesn't like to listen to her own grammatical mistakes. How she feels about my treatment of her native tongue, I don't ask. Instead, I confess my confusion about Lundbeck's famous test and ask for an explanation. Why does my genetic activity proclaim that I am depressed, when I'm not? What is it that's being tested?

"How about a little background first?"

As we move through the wide hallways and pass offices notable for their order and congruent colors, Larsen shares the tale of the pharmaceutical industry's biggest problem. The company executives know they can't go on producing new products that are actually only slightly modified – or "gussied-up" – versions of old friends. Plus, there are lots of unmet needs to be dealt with. But to develop something new or significantly better, the industry's scientists must first uncover and identify a disease's mechanisms. In the psychiatric area, it is just incredibly difficult to get hold of these mechanisms.

"A general problem is that diagnoses are subjective," says Larsen, wringing her hands. When it comes to depression, which is a really big market, they've gradually come to the conclusion that, in reality, the relevant diagnosis – major depressive disorder or MDD, as the psychiatric manuals term it – covers more than one disease. This intuition had been gnawing at the specialists for years. In particular, they know that a third of patients with an MDD diagnosis do not react to antidepressive SSRI drugs, though the rest do. It's just that no one has been able to find the criteria to identify which patients fall into which group in advance.

"We've finally had to admit that there are probably different biological mechanisms involved. There is a difference in the patients' sleep patterns, in which some sleep a lot, while some wake up early in the morning, just as there is a difference in whether the stress hormone cortisol is elevated or lowered. If we separate the different groups on the basis of biology, we also stand to find out whether we should treat them differently. Depressed is not just depressed," Larsen confesses with an audible sigh. "And we would very much like some biomarkers to distinguish them."

She is speaking on behalf of an entire industry. Everyone in "Big Pharma" is working to find biomarkers. Entire conferences address the topic; the market demands it. In a recent report on the future of the pharmaceutical industry, analysts at PricewaterhouseCoopers predict that, by 2020, it will be difficult to sell drugs without, at the same time, having a diagnostic test to verify which precise medicine the individual patient should receive.

"But we have a problem in the psychiatric field. Cancer researchers can easily search for biomarkers in tumors, but we cannot go in and take a biopsy of your brain to look at its biology, right?"

I understand that. On the other hand, I don't quite get how they figure on finding markers for depression and personality disorders in white blood cells, which are the storm troopers of the immune system.

Larsen is eager to illuminate me. "Depressive patients show a number of changes in the immune system, and many believe that the disease may have a component that has to do with inflammation. We may not be able to see the whole picture in the blood, but we may find *something* relevant to measure."

It sounds more like a fishing expedition than a scientific hypothesis, I think.

"Listen, for the last fifty years we have used animal models and tried to extrapolate from animal to human, but this has great difficulties in brain diseases. You can hardly ask a rat whether it feels sad or has hallucinations. It is obvious that we have to begin with patients, but we cannot take brain tissue from living people. On the other hand, we have lots of patient blood from our major clinical trials, and our starting point is that it is better to look at the blood from a patient than at the brain from a rat."

I give in. But I still want an explanation of how in the world you find out which genes you're supposed to measure in the patients' blood. After all, there are over twenty thousand, and who knows which are relevant?

"We have experts for that," says Larsen, abjuring any personal responsibility, "but I know it began with literature surveys."

Specialists in the company's US research department put on their reading glasses and spent six months digging through all the studies that postulated an association between depression and one gene or another. Then, they whittled their way down to a set of twenty-nine genes that seemed likely to play a part. Interestingly enough, these are not genes that have anything to do with the brain's signaling pathways. The receptors and transport proteins we normally hear about are conspicuous by their absence.

"It's a mixed bag," Larsen admits. "There are some genes that regulate the activity in cell nuclei, and a number of genes that are involved in the function of the immune system. But I can't reveal more, because we don't have a finished product yet."

What she can reveal is that, in their initial experiments, they have measured the activity of these twenty-nine genes in blood samples from several hundred carefully chosen people enrolled in clinical tests in Denmark, the United States, and Serbia. These women and men were either healthy controls or people diagnosed with an untreated depression that had lasted for at least three months. Finally, they included some patients diagnosed with either borderline personality disorder or post-traumatic stress disorder. Why those? Borderline patients have fundamental difficulties regulating their emotions and thus display a number of symptoms in common with depression. Post-traumatic stress, on the other hand, is completely different from the other two conditions and could serve as a control for whether the effect the researchers eventually find in the blood is a standard reaction, a signal from the immune system that something unspecified is wrong.

The gene's activity – or *gene expression*, as it is called – is determined by measuring how much RNA is transcribed from the relevant genes. The measurements are next transferred to a computer, where the whole kit and caboodle is combed by various pattern-recognizing algorithms to find characteristic and statistically sustainable patterns. All tasks done by machines.

"The whole project could have ground to a halt right there," Larsen stresses, her voice shivering. "We could have been facing absolute chaos, pointing in every direction. Just imagine it."

But there *are* patterns, patterns that clearly separate the different groups into distinct piles, the healthy from the borderline from the post-traumatic from the depressed. Across the three illnesses, some genes are more or less active in relation to the control condition, but there is a way to distinguish among them. In depression patients, for example, a specific subset of the twenty-nine genes shows less activity. Further analyses exposed possible subgroups among the depressed: patients whose gene expression looked more like each other's than anyone else's. But it's not quite clear if these are biomarkers that could be treated or simply interesting patterns, a description rather than a diagnosis.

"We don't know yet what the test will be able to do," Larsen says. "It's possible that what we're seeing is an acute reaction to the disease itself and that you can identify patients in this way. But, first, we're examining whether the gene activity is normalized when the patients are treated. A major experiment is going on now, where we are testing the twenty-nine genes in patients' blood before, during, and after treatment with two different antidepressive drugs. But it takes time. We're talking about thousands of samples."

In the meantime, they're exploring another possibility, namely that the patterns in the blood are not just acute reactions but an expression of epigenetic changes. And this seems to fit well with my personal test result. I'm *not* depressed but I *have* been.

"It is definitely possible that we have captured an epigenetic change that happened a long time ago but still remains. Maybe, this pattern itself is the expression of what we normally call 'sensitivity' to depression. But if that were the case, it is not necessarily bad news for the test," says Larsen, with a conspicuously broad grin. "It can be used as a handle to find out whether something can be done about the disease and the sensitivity if you can normalize the gene activity in the blood cells."

<hr />

"I DON'T KNOW the details in the test, but from what you're telling me, it sounds likely that we're talking about stable changes," says Moshe Szyf over the phone from his office in Montreal. "That is, not just acute changes but something epigenetic. And, generally speaking, I'm convinced that, when epigenetic changes happen in the brain, it has an effect on the whole system. The immune system speaks directly to the nervous system. So, it is highly likely that changes in the brain will show up in white blood cells."

There is a short pause.

"But it's not other people's work you want me to talk about."

No, most definitely not. I'd spent five minutes waiting on the line of an international call to hear about the potential for the field of epigenetics from one of its leading figures – and one who specializes in the brain. I fully understand that my test from Lundbeck is not his top priority.

"Yes, *now* we get attention," says Szyf. "Some of us have been working with epigenetic modifications of the genome for thirty years, while everyone was totally indifferent."

But now there's enthusiasm, I remind him cautiously.

"Yes, because it's finally clear that epigenetics can provide explanations for quite a few of the questions that have baffled us."

Szyf himself has inserted epigenetic theories into some of those questions. He is convinced that we will find an explanation for the huge health gap between rich and poor in epigenetic effects. A person's socio-economic status puts its mark on the genome, he argues, and this results in conspicuous differences in individual health outcomes within our otherwise affluent Western society. As preliminary evidence, he points out that it is not just that people in lower social classes die earlier, or that they are more frequently struck by the major lifestyle diseases. No, even the course of their illness, and the prognosis, is typically worse than for the economically better off with exactly the same disease. "This difference can hardly be due to the fact that the poor have different genes from the rich. It is so obvious that something epigenetic is going on here, but no one has looked at it," he says.

To test his theory, Szyf and his team recently acquired access to a research goldmine: a huge and representative group of Canadians who have been medically tested and followed ever since they came into the world fifty years ago. So far, doctors have examined white blood cells to study whether there are individual differences in how many and what genes have been epigenetically inactivated by small methyl groups. Szyf is checking to see if any of the observed differences have a connection with the person's socio-economic status early in life.

"And there seems to be something to it!" he says without, however, being able to reveal the final results just yet. I take a chance and ask whether his team might have been able to see epigenetic changes in any interesting individual genes? I'm thinking that the social health differences we typically hear about relate to cancer, diabetes, and cardiovascular diseases. Szyf moans softly.

"Of course, we know about special genes that play a role in those diseases. That speaks for itself."

Yes, but have any of these genes been turned off or are they, perhaps, extra active in either the rich or the poor?

"The patterns we're seeing are spread across the entire genome, and my guess is that there are quite a few genes involved here. We're not getting lost by concentrating on just a few at this point."

I certainly beg his pardon, then.

"But a good deal of the enthusiasm in the field also relates to the fact that no matter where the epigenetic changes are found, it is possible to manipulate them, okay?" he adds, in a slightly more friendly manner.

He's quite right. You just have to pin down the biochemical processes. There is a palette of different enzymes in cells that can deactivate genes by attaching methyl groups, or that can open or close access to DNA strands by manipulating histones. And all these nice enzymes can, in principle, be regulated by well-chosen chemical compounds.

Szyf has demonstrated these mechanisms in action in a group of poor rat babies he allowed to be raised by inattentive mothers. As adults, these badly raised rats were plagued with nervous temperaments because of their geared-up stress reaction, but this could be corrected with one good shot of the chemical compound trichostatin A, or TSA. When the rats received injections of TSA directly into the brain, the chemical erased the epigenetic signature that had branded their conduct as infants. The damaged animals suddenly relaxed and, from then on, displayed completely normal reactions to stress.

"We already know a handful of recognized drugs that work on epigenetic changes," Szyf says. He mentions valproate which, like TSA, inhibits the histone deacetylase and is used to treat depression. "There are also clinical tests going on of similar drugs for the treatment of psychoses. And now, when we are finally, hopefully, seeing some serious investments in this field, a lot of new drugs will be coming out over the next few years."

But how does he imagine the treatment of patients will work in practice? After all, you can hardly stick needles into the brains of mistreated children. Epigenetics generally involves making chemical modifications that might only take place in one particular tissue – blood cells, say, or the brain. If a drug affects the epigenetic patterns in all the cells of the organism, you might well disturb completely healthy gene activity and create unintended side effects. It can't be easy to hit just the right cells.

"Nobody said it was going to be *easy*," Szyf shoots back, somewhat offended. "We have to find out which enzymes are the most important for different tissues and design drugs that are aimed at each enzyme."

But is chemistry even necessary? Couldn't you imagine doing something in a more natural way via simple changes in behavior? If a miserable upbringing and the way people treat us can influence our genome, then can we also influence it the other way? I think again of the tough skin on a sensitive psyche.

"You're probably right," Szyf admits, surprisingly cordial. "I even think that behavioral and psychological interventions will end up proving better than drugs. Because they play directly on the biological mechanisms that we're already dealing with. But I'm not a behaviorist, so don't ask me for specific examples."

I don't ask.

"But the path has been paved for them. If we know the epigenetic signatures and markers – for abused children, for example – we can design behavior therapies, talking therapies, or whatever, and *study* whether they work. Determine whether they remove the markers in question. Today, psychologists are doling out all sorts of therapies without knowing what they actually do to people, but we can test them in the same way they do clinical tests on drugs. We can develop therapies that are tailor-made for different problems."

Still, it's not yet time to forget about the "old" genetics. After all, we know that particular gene variants do have an effect, both physically and psychologically. Should scientists be looking at genetics and epigenetics together, to find out how the two interact? I permit myself to refer to another well-known epigeneticist, Andrew Feinberg of Johns Hopkins University. He has been airing the idea that we might be walking around with gene variants that make it more or less difficult to make certain epigenetic changes.

"Naturally," says Szyf, once again sighing audibly. "I know of research that is right now looking at psychiatric diseases in identical twins and comparing their respective gene variants and the epigenetic patterns each of them has. But you have to understand the field is young. We are at the beginning of something very big, and we don't yet know how big it is."

Actually, I suggest, the really big thing is the discovery of how incredibly dynamic the genome is. This makes me think of an oft-cited quote from 1989, when James Watson, in his usual cocksure way, said to *Time* magazine: "We used to think our fate was in the stars. Now we know, in large measure, our fate is in our genes." Some twenty years later, that sounds like a foolhardy and hopeless simplification. Now, we know that genes are not fate, in the sense that our DNA sequence determines the life we get.

"Of course, our fate is *partially* in our genes, because they lay the foundation for what can happen," Szyf says. "But there is at least as much fate in the way we interact with each other. We have a very significant influence on how the genes we inherit actually work through the way we interact with the world around us. How significant ... research must show."

Szyf has several appointments on his calendar, and I can almost sense the poor people standing in line, waiting for me to get off the phone.

"What we're working on now – depending on how much they invest, of course – will shake up the way people think about biological inheritance." We're moving away from the idea that the foundation of our life is something quite static to the view that it is something very dynamic and malleable. This is a recognition that will spread throughout the culture, and the interesting thing will be to see how it has an effect on the ways we view and use genetic information in the future. That's something to think about."

8

Looking for the new biological man

Finding that "special someone" doesn't happen in a test tube — it's a process . . .
Current scientific knowledge can only take us so far — the rest is up to you.

SCIENTIFICMATCH.COM

There's a pair of white cotton panties in my underwear drawer that I've never worn and never will. It's not because they're a little too big for my taste, but because across their front and the back is printed a long, slightly blurred sequence of letters — A, G, C, T. It's hard to decipher and looks more than anything else like the imprint from a wet newspaper.

But actually it's art. I received these panties from Joe Davis, who, from an office tucked in a corner of the Massachusetts Institute of Technology, concocts art installations involving biological materials such as Petri dishes of genetically manipulated bacteria. Davis hand-printed the panties at home in his kitchen, and the sequence comes from one of his own genes. Not any old gene but one of the so-called human leukocyte antigen, or HLA, genes, which reside on chromosome 6. There is a group of HLA genes, which code for a protein that bristles on the surface of white blood cells, where they play a crucial role in stimulating the reaction of the immune system to infections.

"You have to see this as a commentary on the future and the whole genetic project," Davis explained when he handed me a size medium. Fair enough. But why panties exactly, and why an HLA sequence?

The project harks back to a famous experiment from 1995, in which a group of women turned out to be attracted to the body odor of men on the basis of their HLA genes. The man behind the experiment was a young PhD student, Claus Wedekind, who as a zoologist was well-versed in the world of animals. He noticed that many species – including mice – prefer a partner with certain genetic characteristics: they sniff their way to the set of HLA genes that is the most different from their own. This appeared to have nothing to do with finding *the* ideal mouse mate – a supermouse for all seasons – but with the fact that, for every given mouse, some partners complement its particular genes better than others.

When Wedekind explored the idea further, he realized that, basically, the animals were optimizing the fitness of their offspring. Mates with a big difference in HLA provide the offspring that come out of their union with the greatest possible variation in their HLA genes. This, in turn, bestows the next generation with the most flexible immune response.

And if mice are so 'rational' in their mating preferences, Wedekind reasoned, why not human beings?

He persuaded forty-nine women to stick their noses into a pile of T-shirts from various men – unknown to the women – who had slept in the garment for two nights in a row without using any form of deodorant or other artificial mediation. In other words, pure, unadulterated body odor. Forty-four men participated in the experiment, and their T-shirts were distributed at random, so every woman gave an opinion on six. The women then described how pleasant or unpleasant the odor of these men was in relation to each other and ranked them by personal preference.

Both the men and the women had been gene-tested for three specific HLA genes – A, B, and C – each of which exists in a number of variants. It turned out that the women did not have the same preferences but clearly favored the odor of men whose HLA type was different to their own. The more difference, the more pleasant the smell, the researchers found. At the same time, the preferred smell was consistently the one the women described as being closest to the smell of their present or past partners. The results were so clear that Wedekind concluded: "Our finding shows that some genetically-determined odor components can be important in mate choice."

The findings also showed that birth control pills disturb this scent-guided choice of partner. If the women were on birth control pills, which physiologically imitate a pregnancy, their preferences turned upside down. Then, they preferred men whose HLA type was close to their own.

News of the musty T-shirt experiment traveled the world, providing not only headlines but dubious commentary. Could it really be the case that we human beings, with our highly developed culture and refined ways of life, still have such primitive instincts? We just don't go around sniffing each other – we make social demands of our mates; about appearance, educational attainment, job status, and political views.

Yet a mini-industry of gene research was born. Subsequent experiments from Wedekind and many others have substantiated the fact that odor preference exists and that it applies to both women and men. Field studies indicate that the preference means something in practice. In the United States, Carole Ober of the University of Chicago visited a group of Hutterites, an Anabaptist religious sect known not to use contraception on principle, who agreed to let its married couples be gene-tested. Ober's study of HLA genes found that, to some degree, the Hutterites married following the same rule as mice – the difference between HLA genes in husbands and wives were greater than what would have been expected if the choice of partners were random. Later, a French-British-Chinese team found a correspondence among "regular" white Americans and gave their support to "… the hypothesis that these genes influence mate choice in some human populations."

Today, you can subscribe to dating services based on your HLA genes. First off the mark was Eric Holzle, an out-of-work engineer from Boston, who established a dating agency, ScientificMatch.com. His entrepreneurial brainstorm is that users will buy a test for a number of HLA genes, which Holzle then uses to suggest possible partners who might be a good genetic match. More recently, a quasi-competitor, the Swiss company Gene-Partner, has entered the market, though it does not directly put people together and instead limits its work to testing interested people. As usual, it's quite easy – you get a couple of cotton buds, scrape the inside of your cheek and that of the other party or parties you'd like to have checked, and post them to the company. Later, you receive access to a closed account on the company's website, where you can read at your leisure about how well you supposedly fit together, genetically.

"THIS ALL SOUNDS like some kind of *animal* attraction," says my colleague J, when he hears about GenePartner, and he clearly finds the thought appealing. So appealing, in fact, that he thinks we should both get a test as soon as possible. This does not come from nowhere. For a couple of years, he's been energetically advocating that we produce a baby together. It's not that we should live together or have anything to do with each other in the traditional way – we are both involved with other people – but since his girl-friend can't have children, he thinks the arrangement makes sense. Particularly from a genetic point of view.

"We're both good-looking people, right?" he says, referring mainly to himself. When I shrug at it, he argues further that we complement each other well: "You're gifted in the rational direction, while I'm an excellent example of the artistic, aesthetic type." An ideal combination, in other words. When I'm not convinced, he moves on to the purely physiological advantages.

"My grandmother lived to be over a hundred, and she was fit as a fiddle right up to the end." And then the trump card: "My liver and pancreas are in top form!" Which undeniably says something about a robust physique, when you consider J's consumption of wine. Now, he's trying a scientific argument as a last resort to persuade me: "We can have an HLA test done, which can show whether we should propagate before it's too late for me."

I consent in principle but throw J out of my office so I can call Claus Wedekind. He is no longer a student but a professor at the University of Lausanne, and when he hears of my errand, he releases a heavy sigh.

"I guess I'll have to get used to the fact that I'll never be rid of that old study," he says, explaining that he no longer works with human beings as a model system; he's far more interested in fish. But every week, at least two journalists from somewhere in the world phone him to take a position on T-shirt odors and attraction. This conspicuously polite and soft-spoken man admits that he is a little tired of it. "I refuse to appear on TV; it simply takes too much time. But I'm happy to answer questions over the phone."

I want to know what Wedekind thinks about the research field he helped launch, which – in its limited way – has been commercialized and marketed. Is there even enough data to put a product up for sale? Again, he sighs.

"I believe there must be agreement that there are preferences with respect to body odor that are connected to HLA genes. I also believe that there are preferences with respect to human mating. That can be deduced from my own study, and it has been seen in similar studies repeated in other populations."

That is, there are studies on white Americans. But one of them, released in 2008, found no HLA preferences when it investigated couples of the Yoruba people of Nigeria. And in two other studies conducted in 2000, that tested the HLA genes of three hundred

Japanese couples, the degree of difference between HLA genes in spouses appears to correspond to the population generally.

"The fact that some studies have not found any preferences doesn't speak against the whole idea," Wedekind insists. He mentions that he is inclined toward the hypothesis that the preference for a mate with different HLA genes developed to prevent inbreeding in the small groups in which *Homo sapiens* originally developed and lived.

"We don't know how important it is and how much HLA genes mean for people in relation to other factors. It may also easily be the case that they mean more for some men and women than for others – that there is variation. But if it proves to be something you can use advantageously for dating, I will be proud to have contributed something. But you know what? Call Craig Roberts at the University of Liverpool, he's more up to date on the research."

"THERE ARE LOTS of inconsistencies in the results," says Roberts, an anthropologist, ten minutes later, referring to an article in which he reviews all the experiments and observations in relation to mate choice and biology.

"It's all a big scam," he adds, about the scheme to use the gene as a dating compatibility test. "We're not yet at a stage where we can say that HLA differences have a practical significance. But sex sells, and so does science. People are gullible, and someone is going to make a lot of money off this."

Roberts sounds put out, but he suddenly strikes a plaintive tone. "Maybe I'm just jealous," he says, admitting that he himself tried his hand at the business. A few years ago, he was thinking about starting something along the line of a dating service, and even got a business partner to help realize the project. "But, ultimately, I couldn't make myself do it, because I could see there weren't enough data. I'm too ethical for my own good."

Well, we all have our cross to bear.

"Think of all the noise. GenePartner's website talks about social and biological compatibility, but on the biology end, they don't take into consideration what we know about preferences with respect to faces and how it all goes together with HLA genes."

He has a point. A number of studies have been made on which faces attract us, and when you begin to gene-test the research subjects, it typically proves that the faces people find most attractive belong to people whose HLA genes resemble their own. Roberts, who has done this sort of experiment himself, has a theory that the two preferences guide us down a middle path. If we prefer faces of people with similar HLA genes but scents of people with maximally different HLA genes, we may, on average, choose partners who are in the middle – avoiding the extremes.

"I just don't know where the gold is here with partner preferences," Roberts says, somewhat discouraged. "There are a lot of laboratory experiments and dirty T-shirts, but we know very little about how it works in real life."

I ask what he thinks about the sensational 2006 study from the University of Arizona, in which a group of researchers found that couples who had a large difference in their HLA genes not only reported a better relationship but also had a better sex life than couples with smaller differences. The women in particular reported more orgasms. These same women also cheated less than women who share more HLA genes with their husbands.

"Yes!" Roberts spontaneously replies. "That is an excellent study, and I would love to get the opportunity to repeat it with a different group of people and expand the experiment's design. I believe that you should go beyond sexual satisfaction and relationships and look at whether the offspring, in fact, are healthier and more disease-resistant, if the parents have big differences in their HLA genes. This would address some of the evolutionary theories directly."

In the meantime, however, Roberts is applying for a grant to investigate what people actually think about genes and mate choice and how they would use the information from a dating service like, say, GenePartner.

I'm struck by how old-fashioned all this sounds. So I ask Roberts if he thinks this new genetic approach will change our view of what we seek in a partner. That the venerable goal of producing "good" children is what a relationship is about.

"My gut feeling tells me that nothing will change," he says at first, but then immediately does a half turn.

"Have you seen the clips from *Good Morning America*, where two daters were tested by GenePartner? These two people clearly don't know much about genetics, but they are obviously thinking about a genetically suitable partner."

I promise to watch it, and Roberts' farewell sounds like a paraphrase of Wedekind's: "I feel convinced that an underlying genetic component to the attraction we feel for each other will be found. We just don't have the right take on it yet."

THAT TAKE IS what they're trying to find at Sood-Oberleimsbach, an industrial district near Zürich, where GenePartner has its headquarters in a box-like office building. The company's total manpower proves to be two women: the director, Joelle Apter, and the head of research, Tamara Brown. It was Brown who kindly arranged to expedite the analysis of the romantic compatibility of the HLA genes in me, my boyfriend, and my eager colleague J. She corresponds to none of my expectations for a sharp player at the somewhat dubious end of the gene industry. She has a delicate, pale, and well-scrubbed face that could easily belong to a Renaissance Madonna, but she is dressed in loose jeans and a much-too-roomy grey sweater. And she has a tendency to jabber away candidly about this and that.

"When I was in my twenties and didn't have a boyfriend, I had a plan. If I hadn't found a husband before I hit thirty, I would have a baby by a sperm donor. And I thought 'why not choose a black man?' The mix is so beautiful, don't you think?"

The plan never amounted to anything, it turns out. Above her desk, there are some photographs of a blond – and very attractive – man. It was after catching him via Internet dating that Brown spoke to Apter about whether Wedekind's laboratory findings could, perhaps, be applied in the big, bad world. "We didn't know whether it would work, because it had to do with odors. But we wanted to investigate whether it actually meant something for real couples."

First, in 2003, they established a company, the Swiss Institute for Behavioral Genetics and immediately advertised for couples who had been together for between five and thirty years and who wanted to participate in a research project. It took a long time to find subjects, admits Brown, without, however, revealing how many they actually got hold of. "Sufficiently many to get statistically tenable material," she merely says.

Everyone was HLA-tested and completed an interminably long questionnaire that was supposed to characterize their relationship and how satisfying it was for both parties. Had it been love at first sight or was it a friendship that had grown into something else? How would they describe the state of their sex life, how many children did they have, how many years in between the births?

"We could see that the principle of a preference for differences in HLA genes held up and, at the same time, there were some characteristic patterns in the combinations. It seemed to matter what differences there were between the two. Certain combinations of HLA variants were far more frequent among our couples than people would have expected from the distribution in the population."

When I make reference to the conflicting reports in the literature about the extent to which the HLA preference exists in real life, Brown cocks her head and says "hmm." And when I ask where GenePartner's study is published, I learn that it isn't.

"Not because we couldn't if we wanted to," emphasizes Brown, "but if we tell the whole world what we've done and what we found, we'd have no protection for our business."

The business is an algorithm. From their observations in the volunteers, according to Brown, they were able to calculate how a given combination of HLA variants would fit together – that is, how attracted the partners were to each other and what the chances were that it would hold together over the long run.

"As a trained biologist and ex-researcher, I am quite interested in what makes some combinations better than others. But we can't really go into that right now."

At this point, they have provided tests to well over a thousand couples, most of whom live in the United States; in Europe it is the Swiss who especially seek their services. Of course, some of their clients are people who have just met, but GenePartner also analyzes the genes in established relationships, usually because the parties have heard about the effect, found in Wedekind's original experiment, of the birth control pill on attraction. When they met, the woman was taking the pill, but now she has given them up and the relationship is no longer working so well.

"We don't get to know what happens with them after the test, but it says something about how much faith people have in these genetic factors," says Brown, emphasizing that she's out to substantiate this belief with data. "In fact, we're trying to enter a collaboration with private matchmakers, because they bring people together and get personal feedback after their meetings. With that sort of data, we would be able to document whether it gave a greater success rate to use the HLA test." The first collaboration is in place and will be called DNASoulmate.

While Brown waits for her documentation, business must go on. The company is already developing a new product – an HLA algorithm for homosexual couples. The American dating portal Clickk is offering the service to its gay and lesbian users, and GenePartner is testing away to see whether homosexuals can also use genes to navigate the meat market.

"Our goal is for the HLA test to become a standard in a few years. That people will combine both social and biological compatibility. I believe that genetics must be seen as the last tool to narrow down a field that is already screened for the social aspects we know are important. You know the level of education, interests, goals in life, and all that."

But why stop with HLA genes? There may be other predispositions that could conceivably let us get deep under the skin of our potential partners. I immediately think of the commotion aroused when a Swedish team announced it had identified what the media promptly dubbed an "infidelity gene." The result was packaged in a thin, concise article, published in the well-regarded journal *PNAS*, and it pointed to a genetic variance in men's connection with their partner. Paul Lichtenstein of the Karolinska Institutet hadn't pulled this idea out of thin air. He had investigated the gene for the vasopressin receptor, which is found in the brains of mammals and which, among other things, is cited as a key reason a gerbil lives either monogamously or promiscuously.

The gene, called in everyday language APVR1A, exists in three variants in human beings. Lichtenstein and his colleagues gene-tested 552 pairs of twins and their partners and also gave them questionnaires to characterize their relationship. For example, they were asked to judge the strength of their mutual attachment, which was coded to a scale. And wouldn't you know, gene variant 334, which makes gerbils promiscuous, also proved to have a certain connection to a man's poorer relationship with his partner. Women whose husbands had two copies of 334 were, on average, somewhat less satisfied with their relationship than were other women. And whereas only fifteen percent of men without the variant reported a crisis in their marriage, thirty-four percent of men with two 334 copies did. The latter group also married half as frequently as the study's other men. Of course, write the authors, the results cannot be used to predict the behavior of individuals, but this remark got lost in the media coverage.

On the other side of the table, Brown is intrigued. I neglect to mention I have found a little laboratory in Arizona, Genesis Biolabs, which conducts the test.

I would like to bring more than just sex and infidelity into this conversation. Recently, Neil Risch and Esteban González Burchard of the University of California San Francisco, discovered evidence that two Latin population groups choose partners that have, more or less, the same racial mixture as themselves. The researchers gene-tested Mexican couples in San Francisco and Mexico as well as Puerto Rican couples in New York and Puerto Rico. The couples were compared for hundreds of markers spread throughout the genome. From these, they found that the Mexican spouses resembled each other with respect to their Indian and European roots. The same held true for Puerto Rican spouses – though only with respect to their African and European roots. Now, you might think this just had something to do with coming from the same class or social background, but when the researchers looked into socio-economic conditions, these couldn't explain the results. "People seem to gauge their partners, perhaps on an unconscious level," Burchard said to *New Scientist*.

Brown says that she "certainly sees the possibilities" for testing a wide range of markers. "I think that, at some point, we will expand our service and array of products, but we don't yet know what it will be. But if you ask me how dating sites will look in ten years, there will certainly be a number of genetic services. I can easily imagine that the health-risk profiles from 23andMe and others will be one of the parameters you can assess each other with. And there will certainly be a desire to see the other person's status with respect to one or more genes we know from behavioral genetics."

I wonder whether I would be rejected for my worrying COMT variants. Or my sensitive SERT, for that matter. And what would I find unpromising in a man? Brown claps her hands and pulls me out of my romantic gene reverie, suggesting we move to the computer. "Let's look at your data."

To start, she enters my code and then my boyfriend's into the system. On the screen, a minute scale appears and an arrow points at seventy percent. It is a bit of a mystery to me how you are supposed to interpret the number, but I scan the accompanying, helpful text:

> *This genetic pattern reveals a high level of biological compatibility. Most couples show a corresponding result. This provides a good basis for a very strong and stable long-term relationship. Couples with this genetic pattern often report a high level of physical attraction and passion. However, please be aware that both social and biological compatibility are important for a lasting and satisfying relationship.*

"It's a fine match," concludes Brown. "Most couples in our research project were between sixty percent and eighty percent. You could say you feel seventy percent attracted by him and, in addition, there is a scale for the type of interest that goes with the HLA results. If all your genes are different, you would be romantically very attracted, but if there were more common genes, you could still make an excellent match. In that case, our results indicate that you feel safe with each other, if you know what I mean. And since we saw fewer couples who were maximally different with respect to HLA, I think it may be due to the fact that the attraction is so strong starting out that you ignore social differences and conflicting interests that then later make the relationship fall apart."

It sounds like I should keep the one I've got.

"Yes, he's okay," she answers a little absently. She is already looking at my colleague J, who, I have told her, is a bit of a playboy.

"Whoah, he's almost a perfect match," she says, fingering the screen, where the arrow is hitting eighty percent.

> *This genetic combination is typical for a very satisfying relationship, which also offers a high level of physical attraction. This means that both parties presumably find each other very attractive. This is*

> *important, because it means that the chances for intimacy will not*
> *diminish over time. You will presumably retain a passionate and*
> *highly fulfilling relationship. However, please be aware that both*
> *social and biological compatibility are important for a lasting and*
> *satisfying relationship.*

"Everything looks good here. Aren't you attracted by him?"

Maybe, I am — somewhat — on some level, but the idea is just to have a child together and share custody.

"Yes, well, it would be a good child. Or — at least, there would be a good chance for a successful pregnancy. You have something to think about."

"HOW DID THE tests go?" asks J, when he calls me later in the day from Copenhagen. He's sitting in a café with "a beautiful girl" as he puts it, but still has the energy to think of his possible progeny.

"Did you get the word?"

There's no way around it, and I have to tell him that he looks to be the best of my alternatives.

"I knew it!"

Later in the day, he texts me with an entirely new argument for a quick coupling:

> *Do you realize what this child would do for your book? We could con-*
> *sider it our contribution to literary history.*

CHILDREN. OFFSPRING. YOUNG 'UNS. You can't get around the fact that this is the entire point of this genetic information — its own continuation in new combinations. You can't talk about genetics without getting into children, and you can't talk about today's explosion of knowledge and technological ability without speculating what it means for the children of the future. And, thus, the future of humanity.

So, since I'm trapped in a hotel room in the middle of Zürich, the immaculate city, I do a little research. On my stroll through cyberspace, I happen upon a comment that sums up the day's activities very well. Randall Parker, the man behind the *FuturePundit* blog, writes, "Looking down the line ten to twenty years I expect to see online dating services match people up based on avoidance of shared harmful recessive genes. Searchers for Mr and Mrs Right will get steered toward prospective mates with whom they can pretty safely make babies."

That is not at all improbable. And it reminds me of a conversation I had a long time ago now with an old acquaintance who also happened to be a geneticist. Armand Leroi experiments with mice in his lab at Imperial College in London, but he is happy to talk about human genetics. In his book *Mutants: On Genetic Variety and the Human Body*, he muses about the strange things that can happen when our genes do not behave. Later, in a controversial opinion piece in the *New York Times*, he advocated that characteristic genetic differences make it reasonable to talk about races among human beings. The last time we met, he was talking about calculating the individual's *genetic quotient*.

The conversation took place at a very hip and very noisy restaurant in London's pricey Knightsbridge. Over a bloody steak, Leroi interpreted his ideas about putting numbers on a person's overall genetic health or quality. Somewhat in the style of an intelligence quotient, which provides a general measure of a person's intellectual capacity.

To hear how far he has come I phone, and catch him at home, just returned from a diving vacation in the Red Sea. The conversation takes place over the Internet service Skype and, as we peer into our little web cams, I can see that he hasn't changed much. Slim, appropriately sunburned, and balding in a decorative way. His office, of which I catch glimpses, reflects good taste. There is a

Persian or Afghan rug on the floor and small, exotic wooden sculptures scattered on the available surfaces. The cramped bookcase behind him reaches from floor to ceiling, yet still gives an impression of order and planning.

"Oh, yes, the genetic quotient," replies Leroi, who is apparently just pulling himself together. He lights a cigarette and fills my screen image with smoke.

"And this is something you want to quote me on?"

That was the idea.

"Yes, well, it's been a while since I worked on that project, but it will get going again. At some point."

The inspiration at the time, a few years earlier, came to him, he explains, from a television series about weird mutants – people with a thick fur covering their body, or with minuscule heads, or with a predilection to move on all fours. "I thought at the time that mutations were rare, but it struck me that, as soon as you discuss the topic with people, it turns out that they all have one or more family members with genetic diseases. All families carry a lot of mutations and, even though individual diseases are themselves rare, they collectively represent a massive burden on society. Far greater than you'd think."

He sticks his nose into an enormous coffee mug and types with his left hand: "Have you read my 2006 commentary?"

I confess that it has escaped my attention, and I immediately receive a copy through the ether. It is impossible to restrain a certain astonishment when I see the title: "The Future of Neo-eugenics."

"Let's call a spade a spade," Leroi says coolly. "Modern society has been practicing eugenics a long time. It's already a widespread practice and, believe me, it will become even more widespread."

The spade my friend – quite correctly – calls eugenics is what we normally call "abortion for medical reasons;" – the removal of embryos with genetic or physical abnormalities discovered by genetic tests or ultrasound scans. In 2002, women in the G8 countries decided to abort a fifth of these embryos, which corresponds to approximately forty thousand abortions every year. As a result, the number of inherited illnesses has become rarer, and certain genetic diseases due to mutations in a single gene are, in particular, disappearing. In the United States, very few children with the serious Tay-Sachs neurological syndrome are born, and in Canada, Australia, and Europe, the number of new cases of cystic fibrosis has dropped significantly.

"In other words, eugenics has already had a positive effect on the rate of sickness and child mortality. My opinion piece is an attempt to raise the question: when will it be time to screen all fetuses for all known mutations?"

The question is based on simple calculations. Leroi has checked the frequency of thousands of mutations that we know lead to disease when an embryo inherits a copy from both parents and, with the number that was known in 2006, the risk of finding a monogenic disease in a random embryo was 0.4 per cent, or 1 in 256.

"I think that number is quite high," says Leroi, "and I think it comes close to a frequency that would make widespread screening programs interesting. A risk of 1 in 256 is in the vicinity of the risk that justified screening programs for cervical cancer and breast cancer. And with the many more mutations that are known today, the number would be even higher, if I repeated the same calculation."

What could count against the screening for tens of thousands of mutations is the cost, Leroi admits, but as he himself recalls, "It's falling drastically all the time."

I permit myself to point out the sheer physical and personal costs. If you want to sequence a fetus's complete genome or, at least, gene-test it using a gene chip, you have to undertake an amniocentesis, which carries a risk of miscarriage. But Leroi is thinking along other, more advanced lines. He imagines that in the future we will screen and select fertilized eggs before they even come into contact with a uterus. This process, *pre-implantation diagnostics*, involves taking a single cell from each of the fertilized eggs a couple has produced *in vitro*, and investigating its genome.

"Is it unrealistic to think that, at some point, all children in the industrialized world will be conceived in a test tube?" asks Leroi, focusing his eyes on the web cam. I shrug my shoulders.

"Globally, the use of pre-implantation diagnostics is increasing dramatically," he answers himself. Whereupon he squints and takes a long drag on his cigarette.

"We're not talking about perfection here. You may get eight fertilized eggs, and they will all harbor lots of mutations, but it's about limiting their number and rejecting the worst. As I say, 'we're all mutants, but some are more mutant than others.'"

From this perspective, you could also follow the path suggested by Randall Parker and gene-test future parents. Then, if both carry mutations that might lead to disease if the embryo gets a double dose, you need only resort to glassware and pre-implantation diagnostics to weed out any unlucky fertilized eggs.

This model is precisely what the American firm Counsyl advocates. The company, which was conceived by a couple of students from Stanford University, was launched with great fanfare in 2010, touting what they themselves call a "universal genetic test." With a gene chip, they test mutations that are involved in a little over a hundred genetic diseases, at a cost of $350 a person or $685 for couples. If this rings a bell, it could be because of the DorYeshorim – "upright generation" – project. Founded in the 1980s, in Brooklyn, this worldwide program offers genetic testing for ten recessive diseases that are prevalent among the Orthodox Jewish community of Ashkenazi descent. The project's founders advocate anonymous testing of school-age children; the families never learn the results but receive a PIN. When those families who practice arranged marriages seek a match, they can enter the PIN for the prospective bride and groom into a database, and DorYeshorim will tell them whether the two share unfortunate mutations. If that is the case, no dates need be arranged and no one needs to know exactly why.

For Counsyl, the goal is to make genetic testing as mundane as the home pregnancy test, says the company's director Ramji

Srinivasan. His brother, Balaji, has said that they see themselves as "social entrepreneurs with a mission." The siblings have allied themselves with a high-profile group of advisers, including the psychologist Steven Pinker, who offers an endorsement on the company's website: "Universal genetic testing can drastically reduce the incidence of genetic diseases, and may very well eliminate many of them." Nearby, another Harvard luminary and Counsyl "community advisor," Professor Henry Louis Gates, assures us that he considers the test "... a genuine breakthrough for minority health."

"Yes," says Leroi. "Counsyl has had an extremely positive reception in the media, which is evidence in itself that this is the way the wind is blowing. And their hundred diseases is a good beginning." He punctuates his assessment with a slight sniff. "But you can go much further."

Further, to Leroi, means his notion of a GQ – if not a guarantee to look sharp and live smart, then at least a holistic look at the genome that takes all mutations into consideration. The quotient also encompasses those SNPs that do not necessarily result in a disease but may increase the risk of it. Leroi even wants to include mutations whose effect we have no idea of yet. "The principle is that you measure how many mutations an individual has and, from that, calculate the probability for a number of diseases," he says with the tone of a mathematics lecturer. "Ultimately, you may be able to provide a good estimate of your expected lifespan."

It's simple, in concept. You take a genome, sequence its twenty thousand or so genes, and count up how many mutations they carry – those mutations, that is, that might result in changes in the proteins the genes produce. Changes in proteins are potentially a really bad thing. From the knowledge geneticists already have about an organism's proteins, you can calculate the probability that each of the – presumably thousands – of identified mutations will create a seriously disturbed protein. This is not about small, subtle variations in efficiency but about more or less shattered proteins; and thus mutations thought to be harmful.

"Put it all together and you have the person's total mutational load, or genetic quotient," says Leroi, leaning back in his office chair. "Do you understand?"

I believe so. A few researchers have even begun to do the calculations, it turns out. Carlos D. Bustamente of Stanford University, who also happens to have received a prestigious MacArthur "genius grant," has compared a number of African and European genomes and found that Europeans have far more mutations.

"They don't say in the article that Africans are, therefore, genetically healthier but that, I believe, is the conclusion. *Why* may be more difficult to explain," Leroi riffs.

"It would certainly be interesting to know whether there is a connection between physical quality and mutational load," I reply meekly.

"Yes, yes. And, naturally, you have to study and compare large groups of people to show it. But I feel convinced that a person's mutational load will prove to have a direct connection with that person's general health and also with various other traits. Intelligence and physical beauty, say."

As for himself, Leroi is planning *in silico* experimentation. He wants to run computer calculations using some of the many genomes that are now freely accessible to the public and combine them to produce hypothetical offspring. "Virtual babies!" as he says with the day's first approximation to a smile. "This will show you how much the mutational load varies, if you choose different partners, and thus say something about the extent to which it would be an advantageous choice. Both between possible partners and, eventually, between fertilized eggs that you may produce with a given partner."

He stubs out his cigarette and exhales the last smoke from the side of his mouth.

"The greater the knowledge we acquire about genetics, the more important it will seem to us to know what we are passing on to our children. It will become an almost automatic part of our thinking, because we all want to maximize our options. In the beginning, people will pay for this themselves, and there will be a limited number of users, but at some point the health system will step in, because it is about limiting the burden of disease."

Leroi resolutely empties his coffee mug. He has to teach and doesn't have any more time to chat.

"My students are in their twenties, and I tell them that, when they have children, it will seem entirely normal for them to have their genome saved on their laptops. When it comes to choosing from a handful of test-tube-fertilized eggs, you'll want to choose the one with the fewest harmful mutations. That is, the highest genetic quotient. But, as I said – you don't choose between perfection and the opposite, you choose the least worst."

∞∞

AFTER THIS INSTRUCTIVE conversation, I need a cup of coffee. If I were a smoker, I would probably have grabbed a cigarette with it. That was some heavy stuff to get served up in the middle of the afternoon – *neo-eugenics* – and without a filter.

Ironically enough, it is only the turn of expression that is deeply provocative. For Armand Leroi is completely right that the *practice* is uncontroversial for most people. To get rid of unhealthy and defective children before they become children has long been a standard practice and an option we use at our pleasure. I also think Leroi has it right when he says future generations will find active choices with respect to genetics quite normal, almost natural. We are talking about generations who will grow up with easy, cheap, and direct access to genetic information, just as current generations have grown up with easy and cheap access to all sorts of information technology.

We also know that people want to choose. We know this not just because of the many customers Counsyl has already attracted. We need only look at the development of the old, low-technology business of sperm banks. Here, there is great demand for selection. In the United States, it has become a matter of course for women or couples to choose what traits they want to try to pass on to their children by reviewing pictures and detailed descriptions of possible donors. A British outfit catering to the same urge is the London Sperm Bank, which is the largest operation of its type in the United Kingdom. According to its website, the bank offers a "sperm donor catalogue" from which individuals can chose a donor "... with the most appealing characteristics, just add it to [their] cart and complete the 'shopping' process." That isn't the case everywhere. In Denmark, until recently, opting for a sperm donor was a blind choice. A pig in a poke, where, when you decided to use a sperm donor, you knew even less about your child's father than if you picked someone up at a bar. According to the country's rules, it is still not legal to give people a choice, but this looks likely to change under the European Union regime. As Peter Bower, director of the European Sperm Bank in Copenhagen, put it to me, "This is where you see market growth."

But it's one thing to pick from among volunteers who have signed away their sperm and another to poke at defenseless embryos to ensure their GQ is up to scratch. The dawn of *Homo sapiens 2.0* will not happen without resistance and debate. In 2010, for the tenth anniversary of the mapping of the human genome, the journal *Nature* asked a number of prominent researchers to predict real developments over the next ten years. David Goldstein, a geneticist at Duke University, pointed to the screening of fetuses. "The identification of significant risk factors for disease is bound to substantially increase interest in embryonic and other screening programs," he writes. "Society has largely already accepted this principle for mutations that lead inevitably to serious health conditions. Will we be so accommodating of those who want to screen out embryos that carry, say, a twentyfold increased risk for a serious but unspecified psychiatric disease?"

Good question, Dr. Goldstein. It raises the classic debate about the extent to which boundaries should be drawn and rules should be drafted governing what future parents may be allowed to decide. What diseases make it okay to choose an abortion? Which are too minor? Are there traits beyond actual diseases that you may select *for*?

At heart, these issues come down to where the power over our reproduction rights should lie, and where the balance between consideration for the individual and society should rest. The state has always tried to control human reproduction. In certain places and at certain times, it has done so quite bluntly by forbidding abortion and access to contraception, or by ordering the sterilization of selected groups. Today, we may be able to empathize with a couple who want to abort a fetus with genetic defects that will cause great suffering and pain, perhaps even premature death. But what about the less serious conditions – where do we draw the line?

For the time being, the debate has mostly focused on late abortion. For example, in 2008, the Danish Abortion Board, which must grant permission in all cases of late abortion, refused to allow a woman named Julie Rask Larsen to abort a fetus at week twenty, when a scan showed a missing left forearm. The board believed this was a handicap that both mother and child could live with. Rask Larsen did not agree. So, she went to England to receive her abortion.

As people begin to consider the ramifications of our new genetic prospects, they may imagine a brutal society that embraces the crude and narrow view that humanity only has room for the perfect – a real-world *Gattaca*. More simply, parents may start to acquire an "erroneous" relationship with their offspring, viewing children as something they can, like any other piece of technological merchandise, order with the specifications they desire.

A glimpse of this future came in 2009, when doctors at University College, London announced the birth of the world's first child "designed" to be free of mutations in the BRCA-1 gene. The father of the child had seen most of his close female relatives succumb to breast and ovarian cancers because of mutations in the BRCA-1 gene and, with his wife, requested *in vitro* fertilization and screening of the fertilized eggs. From these eggs, the doctors selected one without BRCA-1 mutations. *Voilà* – the end of eternal angst and annual checkups with breast surgeons into the next generation. What could be a better motive for choosing your child's genetics?

But the critics were many. They argued that it was insane to abort innocent fetuses with a predisposition for disease that might never come to fruition or might do so only late in life. Especially diseases that can be treated – and perhaps, within a lifetime, cured.

Such outcries were turned on their heads by the British bioethicist Jacob Appel, who calls for mandatory testing of fetuses for serious genetic defects, including carcinogenic mutations in the BRCA genes. "Mandatory genetic testing isn't eugenics, it's smart science," reads the confrontational headline of his article on the subject. Appel offers an executive summary of this ingenuity: first, the children who are not born are spared suffering from the disease and painful treatments. Second, there are great societal savings in avoiding various expensive cures. And if you look at the problem realistically, he concludes, it is the equivalent of child abuse to choose to give birth to sick and handicapped children, when technology lets us avoid it.

This is where he lost a lot of people. It was just too cruel, even for a philosopher.

As Appel points out, however, Western societies today often put the welfare of the child ahead of the parents' wishes. We don't allow parents who are Jehovah's Witnesses to deny their children necessary blood transfusions. Nor is it legal for members of the Christian Scientist sect to insist on praying for their children's health instead of allowing them to be treated with antibiotics for life-threatening infections. "Child welfare laws certainly prevent a mother from intentionally exposing her daughter to an environmental toxin that produces an 80% risk of future cancer," notes Appel. Why then should the same mother be allowed, in full knowledge, to give her daughter a carcinogenic BRCA mutation?

In the wake of Appel's radical stance, the editor of *New Scientist*, Michael LePage, tried to find some steady ground. He disagreed that gene testing should be mandatory. Instead, society should make genetic screening available to future parents, who can then choose whether to avail themselves of it. In LePage's view, a couple shouldn't have to go to a commercial enterprise such as Counsyl and empty their wallet to access its genetic riches. This is a public good, and the public should provide it. Those couples who get screened and discover they both carry the same mutations may then be offered the option of testing eggs fertilized *in vitro* and selecting the healthy ones.

The central argument of both men is something we are not used to thinking about – namely, the sin of omission. Usually, bioethics is about forbidding or limiting the use of a new technology, and people only rarely ask whether it might be unethical not to use it. But when we have the means to avoid suffering in unborn children and do not exploit it, that is akin to standing around and watching while someone dies of a disease that can be treated by the technology available.

It's at this stage that the worriers truly raise the specter of a *Gattaca*-like existence, in which it becomes socially unacceptable to give birth to a child with a serious disease or physical handicap; socially unacceptable to choose consciously not to be screened. Wouldn't the inevitable consequence be a miserable human community in which the imperfect and the handicapped are poorly treated?

So far, the nightmare scenario has not happened. In practice, society does not treat the handicapped worse than it did before amniocentesis, gene tests, and legal abortion existed. On the contrary. Most would agree that both the treatment of and general attitude toward the disabled are a great deal better than they have been. It might well be that future parents who insist on their right to a sick child will encounter condemnation here and there, just as smokers and obese people presently face societal disapproval. But the condemnation of other people is something humans have always been forced to contend with – this is how morality expresses itself in everyday life. The real challenge is to secure equal opportunity for all citizens, regardless of their genetic profile.

Still, the question remains: if it is ethically responsible to have an abortion to prevent suffering, who is to define what suffering qualifies when the measure is a genetic test? At the beginning of 2010, Britain's Human Fertilisation and Embryology Authority made a first stab at solving the dilemma, producing a list of more than a hundred genetic illnesses for which the country's doctors could offer to screen fetuses. A glance down the list reveals that some are far from life-threatening, such as certain forms of congenital blindness and deafness. There are also diseases that can be expressed in varying degrees, including thalassemia, which, in addition to anemia, can produce pain and much discomfort. But you can also live perfectly well with some forms of thalassemia, as does the tennis player, Pete Sampras.

SO, HOW DO you defend aborting fetuses for this sort of minor infraction? Or something even more innocuous – sex, for example? Several companies, including the Texas-based IntelliGender and the Boston-based Acu-Gen Biolab, have marketed at-home tests that reveal a fetus's sex after ten weeks of pregnancy. If parents can identify sex within the window when an abortion is solely their decision, what would stop them from making such a choice? Ultimately, the "right" to abortion is a political position. Once a country has decided to make abortions available, it has, in principle, opened the doors for a woman to get rid of any fetus she doesn't want. It may be that she does not have time for a child given the demands of work and life; it may be that she has been abused by the child's father and does not want to bear his offspring. These are perfectly legitimate grounds for making this personal choice, because abortion is about a woman's right to make decisions about her own body, under the law. Considered this way, you can see why it is difficult to establish rules that forbid a woman from having an abortion on genetic grounds once a society has made the fundamental decision to recognize the right to either abortion or gene testing. Ethically, it makes no sense.

But even if we come to terms with the possibilities of the present, there is always the future to fret about. When discussing whether or how to ensure we have healthy children, it seems we inevitably wind up debating the prospects for "designer" children, the notion of the gene-manipulated *wunderkind*. Most people shudder at this, which makes it a favorite dystopia among philosophers. Beyond the realm of Hollywood thrillers, the American intellectual Francis Fukuyama, for example, is renowned for vision of a "posthuman" future, in which human beings have taken evolution into their own hands and altered human nature itself. For the worse, it is to be understood, and to a point where liberal democracy is itself in danger. In his horror script, humanity is permanently divided into an underclass of poor "naturals" and an upper class of rich, superior, gene-manipulated individuals – the "created" class, you could call it. And if you believe the dystopians, this new *übermenschen* will consist of blond, blue-eyed, athletically gifted specimens who work as brain surgeons or theoretical physicists by day and are piano virtuosi by night. Of course, such a society of upgraded people is a utopia according to the transhumanists, a collection of philosophers, technologists, and general enthusiasts who believe we should improve humankind by any means available. Unfortunately, the resolution to this debate cannot be "To each his own."

It is high time to dispel this talk. There is no magic gene dust you can sprinkle over Mr. and Mrs. Smith's fertilized eggs to upgrade their future children. The science tells us very clearly that this is naïve and unrealistic. It is based on the outdated presumption that there are identifiable genes that make the difference between tall and short, gorgeous and plain, highly gifted and mediocre. There aren't. Just as there is no such thing as genetic or biological perfection – Armand Leroi's formulation that the best is, in reality, "the least worst" is exactly right.

———

HOW GENETIC KNOWLEDGE will affect us after we are born is quite another discussion. If all our genomes are published and publicly accessible, far more than physical disabilities will be revealed – and not necessarily to the people you'd like. Things like your psychological dispositions.

This prospect inspired Ilina Singh and Nikolas Rose to pose some thorny questions in a splashy opinion piece in *Nature*. "Will 'risk' and 'potential' eventually dominate ideas of personal identity?" they ask. "And will these ideas become institutionalized within education, law and policy?"

These biologically conscious social scientists, based at the London School of Economics, wonder how genetic predictions will affect our lives if those predictions are with us from birth. How will knowledge of the risks and potentials in a newborn's gene profile influence her parents? Will it color their view of their child, change the way they treat the apple of their eye? And what will it do to the child's view of himself that he is labeled from the outset as being particularly sensitive, particularly robust, or placed in some other category of genetic "health?"

With such hypothetical questions, Singh and Rose are breathing life into the idea that purely statistical dispositions may become self-fulfilling prophecies. Simply by reading about a tendency within their genes, a person may begin to behave in ways that conform with their likely expressions and effects. You can easily envision ugly scenarios in which overly heavy-handed expectations crush or lead astray unfortified souls – and not just at Chinese summer camps.

On the other hand, you can also imagine how some genetic alarm bells may trigger an effort to equip more fragile temperaments with vital defenses; a helping hand given to sensitive souls, because genetics tell us that they are especially susceptible to positive influences.

These perspectives reach far beyond the family and its privileged, sheltered corner of the private sphere. Now that this genetic knowledge exists, how long will it be before politicians really join the game and try to base policies on it? There are those who believe we can and should.

When I was on my way to Daniel Weinberger's office at the National Institutes of Health, he pressed an article into my hands, saying *this was something I had to read*. "This is one of the most well-run studies I've seen in years. It's simply beautiful," he gushed. If the topic were not so controversial, he noted, the article would probably have been published in one of the major journals.

Back at home, I turn the pages guiltily. Sure enough, the research team from the University of Georgia has ventured into territory where few people have dared. They went way out into the backwaters of the state and canvassed a poor black community. They then selected 641 families, all of which had an eleven-year-old child. Children on the cusp of puberty, with all the risks that implies for getting involved with alcohol, drugs, and sex – "risk behavior," as professionals call it. The researchers wanted to gene-test and observe the children until they turned fourteen and investigate two things: first, whether children with the short SERT variant are particularly predisposed to risk behavior; and second, whether there is a genetic difference in how children react if those around them are trying to keep them from getting into trouble.

Once the children were gene-tested, they were divided into two groups. In one of the groups, the kids were allowed to go their own way, while those in the other, together with the whole family, were entered into a program called the "Strong African American Families Program," or SAAF. This support program teaches parents how to participate in their children's lives and, particularly, how to set limits for them. Psychologists know this program works. Statistically speaking, the effort has a positive effect on children's risk behavior.

Over the next few years, the children carrying the short version of SERT, who were also in the group left to themselves, began to smoke, drink alcohol, and have sex. And they did so at twice the frequency of the free-for-all kids who had two copies of the long SERT version. It looked as though these children were genetically predisposed to throw themselves into risk behavior, an observation that corresponded neatly to the researchers' presumptions.

Yet, a far more sensational finding awaited, and it had to do with the group that had been enrolled in SAAF. The program had a considerable preventative effect on the "genetically disadvantaged" individuals but only a slight effect on the rest. For both SAAF groups, however, the frequency of risk behavior hovered around the same level as for the children with two long SERT variants who did not participate in the support program. Intervention clearly worked best for the children who were particularly genetically sensitive. Or, rather, susceptible.

Undoubtedly, there are those who will cross themselves, get nervous palpitations, and mumble incantations about stigmatization and social disadvantage. But it could also be said that this research speaks directly against the fear that behavioral genetics will only point out "bad" genes and be the cause for labeling some as hopelessly biologically inferior. On the contrary, this study shows that social initiatives pay off where the problems are most dire. It is not about young children getting gene-tested and stamped as more or less suitable for help. Rather, the very knowledge we get from new genetic studies can shake up our understanding and change policy – for the good.

A small group of social scientists is beginning to realize these possibilities. One of the more ambitious is the criminologist and author Nicole Rafter, whose book *The Criminal Brain* was published in 2008. In the book, she recounts the dubious history of biological criminology, its mistakes, and its hopelessly unscientific basis. But, in a surprise twist, Rafter concludes that biological studies in the modern context may – like the Georgia study – be good. "Today's biocriminologies … are not more of the same," she underlines. Behavioral genetics turns the genetic determinism of the past completely upside down, with huge implications for research, treatment, policy, and the relationship between researchers.

Rafter speaks warmly of a new "biosocial" thinking, a way of thinking that marries sociological and biological understandings of why people behave the way they do under different circumstances. Biologists must definitively refute the earlier medical model of viewing behaviors – including criminality – as healthy or sick, normal or abnormal. Social scientists, for their part, must open-mindedly incorporate biological knowledge into their theories. If that happens, it may be the most effective way to create the kind of programs that address criminality by treating social ills. As Rafter puts it at the end of her book: "I want to enlist modern genetics in progressive social change."

Can genetics really bear this?

Possibly. But it will require a brutally honest liquidation of some hardy old myths. And this requires that more people gain a realistic view of what genes are and what they can do. This will be no easy task, but there are some pioneers blazing the way.

One example comes from psychiatry, where practitioners are now in the process of tossing out the concept of "risk" genes in favor of genetically determined susceptibility. The talk about "orchid" children and "dandelion" children is not just poetic, it is an important change in consciousness: the focus shifts from the risk of an unfortunate outcome to the potential for a good one. And this is a potential that is not determined by the genome itself but by external circumstances. There is no genetic determinism here.

Another necessary readjustment centers on the idea that the genome is something static. Many people have a sense that, because our genes can't be changed, we are in some way or other trapped into a biological straightjacket. As we identify more epigenetic mechanisms, we can see that this is not the case. With its quirky switches that turn gene activity off and on, or turn its volume up and down, the genome is incredibly dynamic. And though we are just beginning to scratch the surface of genetic plasticity, we can already see how information that is in itself immutable is always subject to interpretation. By different tissues, and by all sorts of environmental and other external circumstances. This interpretation can make for colossal differences. The upshot is that the most effective way to shape human beings is not to change the genes themselves but to change what we subject our genes to.

In fact, the rise of epigenetics will undoubtedly lead to a renewed and, hopefully, intense interest in the environment in the broadest possible sense. All the indicators point toward the fact that genetic research is increasingly becoming a holistic investigation of the eternal game of ping pong that the genome, the organism, and the rest of the universe are playing. In other words, it is revealing the dynamism and complexity that is the fundamental condition of biology.

DYNAMISM AND COMPLEXITY are key to these shifts. I predict that, in the future, a third watchword might prove to be *diversity* – in the sense of genetic and, thus, biological and behavioral diversity.

Why that? Because, in the near future, a degree of diversity we never dreamed of is going to come crashing in on us. Researchers will have thousands, and soon millions, of individual genomes to play with, and the exercise will provide fascinating insights into how different our genomes are, how they are different, and what the difference means. We've already had a foretaste of this new menu of diversity, with projects that map and compare broad ethnic groups – or races, if you will. But before long we will presumably see genomes from many more specific groups: the San Bushmen and the Pygmies, the Inuit and the Australian aborigines, and everyone in between. Within each of these groups, diversity will unfold in ever more genetic detail and, when it comes down to it, this may be a welcome corrective to the earlier insistence on what was common among humans.

This will rewrite the mantra of genetic research. In toasts and official releases, it has been said that what is interesting about study-ing different genomes is discovering how *uniform* they are. This has forced the genome into a political role, to serve as a fraternization force across cultural, historical, and social differences. But at this point in the history of science, don't we know perfectly well that we are and will remain one species, regardless of our genetic differ-ence? What is really exciting are those things that, despite every-thing, make us different from each other in so many ways.

The geneticist Bruce Lahn, of the University of Chicago, and the economist Lanny Ebenstein, of the University of California, Santa Barbara, say we need to prepare ourselves for this biological future. It may well be, the two investigators point out in a commen-tary in the journal *Nature*, that research will indeed expose differ-ences we don't care for, perhaps a biological difference that is politically repulsive. According to Lahn and Ebenstein, we need "… a moral response to this question that is robust irrespective of what research uncovers about human diversity."

This is courageous. For they are not only up against historical racial ideology and its attached idiocy but also the more recent debates about the extent to which, and in the given case why, there are average differences in intelligence across different groups. The last flashpoint burst on the scene in 1995, when the book *The Bell Curve* pointed to a so-called "intelligence gap" – and became a best-seller. Armed with many years of measurements from a number of US ethnic groups, its authors Richard Herrnstein and Charles Murray concluded that there appeared to be a typical pattern of bell curves describing the distribution of intelligence within a population. They plotted the curve for Caucasians in the middle of the field, while the one for blacks was lower and the one for Asians higher. Not a popular finding. There wasn't much to call into question in the curves and the measurements taken from the data; instead the outcry swirled around the extent to which the differences are due to genes or reflect purely environmental effects.

For all the heat the book generated, it had one chilling result. A wing of researchers, led by the British neurobiologist, Steven Rose, took the position that the best way to avoid the problem of politics getting into genetics is to refuse to investigate differences in intelligence. Nothing will come from it but discrimination, they maintain.

Lahn and Ebenstein stand in opposition to that position. They argue that a thorough exploration of genetic diversity – whatever it may bring to light – can act as a medicine against discrimination, simply because genetics will make it obvious that it is impossible, if not to say ridiculous, to rank groups or individuals on some one-dimensional scale. Genetic diversity contributes to variation across domains – both physical and mental – and there is no single measurable trait, such as IQ, that says anything exhaustive about an individual's total mental capacity. "We argue for the moral position that genetic diversity, from within or among groups, should be embraced and celebrated as one of humanity's chief assets," they insist.

And we love diversity, right? In nearly every possible situation, difference is elevated as a value. Modern society cultivates and celebrates cultural diversity, and many of us react to globalization's threat of general homogenization by appreciating what is distinct or unique. That which we don't already know. And as far as nature is concerned, diversity is king. Monoculture is the great sin of industrialized agriculture, and environmentalists are fighting a bitter struggle for biodiversity, to save obscure toads, unusual corals, unseen birds, undiscovered beetles, and other living things battling the threat of extinction.

In fact, biodiversity is well on its way toward becoming the next great issue in environmentalism. So why not cultivate and protect our own biological diversity?

The consciousness that we are a species with characteristic genetic differences and physical variation could be the hook that finally gets us to worry about threatened peoples; human populations whose way of life, cultural peculiarities, and language are on the verge of extinction – along with their special genetic composition. It might be the San Bushmen of the Kalahari, the marginalized Udege of the Russian taiga, or the Amazon's Akuntsu – of whom only six survive. To mention just a few.

OUR FASCINATION WITH genetic diversity may also offer an entry to a better understanding of human cultural and intellectual diversity. At least, the psychologists Matthew Lieberman and Baldwin Way, of the University of California Los Angeles, have put forward some ideas about how genetic differences between ethnic groups may help determine where a given culture puts down roots in the world.

Lieberman and Way have contrasted Asian culture and Western culture. They put aside the subtle differences between Japanese and Vietnamese, or French and British and concentrate on the characteristic difference that Heejung Kim studied: namely, that Eastern Asian culture is collectivist, while Western culture is individualist. For decades, anthropologists have studied and described how this is expressed, but these two psychologists decided to ask *why*. Is it entirely random, or might there be a biological basis? Their hypothesis is that social sensitivity marks the cultural divide.

That idea is grating to many ears, especially those adorning people inclined toward the humanities. But Lieberman and Way have some interesting observations to hang their hat on.

Specifically, they investigated whether sensitive – or susceptible – variants of a number of specific genes appear at different frequencies in the two lumped cultures. They reviewed studies of social sensitivity in relation to three selected genes: the by-now-familiar MAOA and SERT and the gene for a receptor found in the brain that is activated by opioids (opium and morphine derivatives). These three genes are all found in a variant that has been proven to increase sensitivity to social stress, but which also provides a high susceptibility to the beneficial effect of an environment where there is a high degree of social support.

Lieberman and Way found that the frequency of the sensitive variants of all three genes is between two and three times as high in Asian peoples as it is in Caucasians. The psychologists suggest that this means more Asians thrive when they receive a high degree of social support and positive social relationships – which, as it happens, is best achieved in a collectivist culture, where people are embedded in a strong social network. This might explain why, for instance, the teaching of the Chinese philosopher Confucius, that the family and the group is the most important thing an individual must take into consideration, has been embraced throughout Asia. In contrast, fewer Caucasians are particularly sensitive to social rejection and exclusion, which may be why ideas about the individual's need to take precedence over the community have been more popular in Europe and the rest of the Western world. As Liebermann sums it up, "When enough brains are predisposed to find the same idea compelling, it is likely to stick around for quite some time."

This is a far cry from traditional cultural research. You can imagine the vicious accusations about genetic determinism and reductionism that will be aired in the professional journals. But the question is whether this and similar research into the genetics of culture will herald a shift in the way we think about humanity, a shift that can be felt in the social scientist James Fowler's call for "a new science of human nature." More to the point, Fowler says that no human science can explain human behavior and culture without integrating human biology, from genes to brain function.

---- ∞ ----

THIS RESEARCH, AND the researchers behind it, are moving us toward a genuinely biological view of humanity. A view that takes as its starting point what the creature *Homo sapiens* is, and the understanding of which is based on both a knowledge of evolution, genetics, and brain physiology *and* of culture and history. This is not about biological man standing in opposition to cultural man. Instead, the goal is to find a more comprehensive familiarity that integrates all the products of human behavior and ideas – politics, culture, music, poetry – and considers them in a biological context. And vice versa.

If there is anything that can drive this transformation beyond the sanctuary of academia, it is personal genetics, quite simply because the phenomenon is a teaching device for the individual that works on the individual's premises. Even now, as genetic tests can be bought in supermarkets and pharmacies, tens of thousands of people around the world are getting acquainted with their genetic information by *using* it. This is decisive. Because it is only when you get the information in your hands and, so to speak, under your skin, that you really experience and understand its significance.

There is no doubt that personal access and the personal approach to genetic information is here to stay. At the moment, consumer genetics is portrayed as a panacea for the plague of diseases, a cornucopia of health and prevention – with the Holy Grail being the advent of personalized medicine, tailor-made for your individual genes. And while illness is an important matter of genetics, it is far from the most important.

Deep down, genetic consciousness is the remaking of consciousness itself. Today, we can check out a million SNPs; tomorrow our whole genome, and the day after that, our genome and its epigenetic changes as they play across the body's tissues and organs.

Eventually, we won't be able to conceive of our self without this information. It will seep into our knowledge of who we are — as a species among other species, and as individual beings.

The Canadian philosopher, Ian Hacking, frets about what will happen when society, as a whole, buys into the new biological view of humanity. "I'm a conservative reactionary," he admits in an essay on consumer genetics and identity. "I know that although my genetic inheritance constrains my possibilities of action and choice, I do not believe it is my essence or constitutes my identity … How long will it take before this attitude becomes extinct? We know that the genomic revolution will radically change the material conditions of life for soon-to-be-born generations … What will be the conception of self for those people soon to come?"

I think he is worrying unnecessarily. As I can bear witness, seeking and finding out what your DNA says inevitably has the wonderful effect of raising a lot of questions, personal and profound.

Who are we? Where do we come from? What is our place in the world? Where are we going? What do we want? In the past, these questions were traditionally relegated to the "spiritual" realm, but I believe they will be better answered by digging deeper into our physical reality.

Now that I have looked into my genes, the result is not a simplified self-image. On the contrary. It is rather that I'm experiencing more facets and nuances in my life. It is far more satisfying to be able to interpret myself as both a biological and a social being. My genes are *not* fate but cards I've been dealt, and some of those cards give me a certain amount of latitude in playing the game of life. Or, to turn another phrase, my genome is not a straitjacket but a soft sweater to fill and shape, to snuggle up and stretch out in. It is information I can work with and around, information that can grant me greater freedom to shape my life and my essence. It is also information that can, in its way, ease my existential burden. It tells me that I am not totally free, but neither am I completely responsible for who I am and what I have ultimately become.

So who am I?

I am what I *do* with this beautiful information that has flowed through millions of years through billions of organisms and has, now, finally been entrusted to me.

Notes

Prologue: My Accidental Biology

8   "the limitations are sociological": Henderson, Mark. 2009. "Genetic Mapping of Babies by 2019 Will Transform Preventive Medicine." *Times* (9 February).

1   Casual About Our Codons

12   "people who ... deal with black employees": Hunt-Grubbe, Charlotte. 2007. "The Elementary DNA of Dr Watson." *Sunday Times* (14 October).

12   "The most unpleasant human being": Conniff, Richard. 2006. "Discover Interview: E.O. Wilson." *Discover* magazine (24 June).

14   "... a historic defeat": Wade, Nicholas. 2009. "Hoopla, and Disappointment, in Schizophrenia Research." *New York Times*, TierneyLab Blog (1 July). http://tierneylab.blogs.nytimes.com/ 2009/07/01/hoopla-and-disappointment-in-schizophrenia-research

19   "... we had found the secret": Watson, James D. 1968. *The Double Helix: A Personal Account of the Discovery of the Structure of DNA*. New York: Touchstone.

27   "The most wondrous map": Clinton, Bill. 2000. White House press conference (26 June). Transcript available at www.dnalc.org/view/ 15073-Completion-of-a-draft-of-the-human-genome-Bill-Clinton.html

28   "Hitler": Shreeve, James. 2004. *The Genome War: How Craig Venter Tried to Capture the Code of Life and Save the World*. New York: Alfred A. Knopf.

32   a team led by Robert J. Klein: Klein, Robert J. 2005. "Complement Factor

H Polymorphism in Age-related Macular Degeneration." *Science* 308 (5720): 385–9.

32   McGill University tried to locate the genetic factors: Sladek, Robert, et al. 2007. "A Genome-wide Association Study Identifies Novel Risk Loci for Type 2 Diabetes." *Nature* 445: 881–5.

32   Britain's Wellcome Trust backed similar studies: Wellcome Trust Case Control Consortium. 2007. "Genome-wide Association Study of 14,000 Cases of Seven Common Diseases and 3,000 Shared Controls." *Nature* 447(7145): 661–78.

33   "We have entered the era": Pinker, Steven. 2009. "My Genome, My Self." *New York Times Magazine* (7 January).

2 Blood Kin

36   "I note to myself": Homes, A.M. 2007. *The Mistress's Daughter: A Memoir.* New York: Viking.

38   "At first, it made me scared": Padawer, Ruth. 2009. "Who Knew I Was Not the Father?" *New York Times Magazine* (17 November).

41   As … Gina Paige told the BBC: Goffe, Leslie. 2009. "Americans Seek Their African Roots." *BBC Focus on Africa* (29 June). http://news.bbc.co.uk/1/hi/8117258.stm

43   "In this future-obsessed era, it is important to seize a snapshot": Indigenous Peoples Council on Biocolonialism. "Genographic Project Director Spencer Wells, IBM Lead Scientist Ajay Royyuru Answer Questions about the Project." http://www.ipcb.org/issues/human_genetics/htmls/geno_q&a.html

43   the past's great sailors, the Phoenicians: Zalloua, Pierre A., et al. 2008a. "Identifying Genetic Traces of Historical Expansions: Phoenician Footprints in the Mediterranean." *American Journal of Human Genetics* 83(5): 633–42.

43   the genome of today's Lebanese population: Zalloua, Pierre A., et al. 2008b. "Y-chromosomal Diversity in Lebanon Is Structured by Recent Historical Events." *American Journal of Human Genetics* 82(4): 873.

46   the most widespread haplogroup among European men: Balaresque, Patricia, et al. 2010. "A Predominantly Neolithic Origin for European Paternal Lineages." *PLoS Biology* 8(1): 1–9.

49   Y chromosomes in … a bunch of other Kohanim: Hammer, Michael, et al. "Y Chromosomes of Jewish Priests." *Nature* 385(6611): 32.

50   tracked down a man she suspected of being a descendant: Harmon, Amy. 2007. "Stalking Strangers' DNA to Fill in the Family Tree." *New York Times* (2 April).

3 Honoring My Snips, in Sickness and in Health

73   rs9642880 apparently increases the risk of that cancer: Kiemeney, Lambertus A.A., et al. 2008. "Sequence Variant on 8q24 Confers Susceptibility to Urinary Bladder Cancer." *Nature Genetics* 40(11): 1307–11.

82   a group of American geneticists warned: Hunter, David J., et al. 2008. "Letting the Genome out of the Bottle: Will We Get Our Wish?" *New England Journal of Medicine* 358: 105–7.

87   deCODEme found that one such variant … if it comes from your father: Kong, Augustine, et al. 2009. "Parental Origin of Sequence Variants Associated with Complex Diseases." *Nature* 462: 868–74.

91   "… who – in their right mind": Long, Camilla. 2010. "When DNA means Do Not Ask." *Sunday Times* (14 March).

4 The Research Revolutionaries

114   variants in a gene by the name of ZFHX3: Gudbjartsson, Daniel F., et al. 2009. "A Sequence Variant in ZFHX3 on 16q22 Associates with Atrial Fibrillation and Ischemic Stroke." *Nature Genetics* 41(8): 876–8.

116   "Humans are really good at … a bit of knowledge": Check Hayden, Erika. 2010. "The Human Genome at Ten." *Nature* 464(1): 664–7.

118   tested relatives of Alzheimer's patients for the ApoE4 variant: Green, Robert C., et al. 2009. "Disclosure of *APOE* Genotype for Risk of Alzheimer's Disease." *New England Journal of Medicine* 361: 245–54.

119   followed … people who bought a SNP-based gene profile: Bloss, Cinnamon S., et al. 2011. "Effect of Direct-to-Consumer Genomewide Profiling to Assess Disease."*New England Journal of Medicine* 359: 2192–3.

119   "Up until now there's been lots of speculation": Tierney, John. 2011. "Heavy Doses of DNA Data, With Few Side Effects." *New York Times* (17 January).

120   "I believe the most important ethical, legal": McCarty, Catherine A. 2009. "To Share or Not to Share: That Is the Question." *Genomics Law Report* (15 Oct). www.genomicslawreport.com

123   "This sense of monopoly prevents": Davies, Kevin. 2009. "Linda Avey on an Alzheimer's Brainstorm." Bio-ITWorld.com (24 November). www.bio-itworld.com/news/11/24/09/Linda-Avey-Alzheimers-brainstorm.html

125   "… cool it is to be able to give something back": McCabe, Jen. 2009. "Exploring the 'Me-ome' 23andMe Research Revolution Results." *Health Management Rx* Blog (23 July). http://hmrx.posterous.com/ jensmccabe-exploring-the-me-ome-23a

128   "I'm a believer": MacArthur, Daniel. 2009. "23andMe Launches New Effort

to Recruit Patients for Disease Gene Studies." *Genetic Future* Blog (7 July). http://scienceblogs.com/geneticfuture/2009/07/23andme_launches_new_effort_to.php

128   The first results of the research: Eriksson, Nicholas, et al. 2010. "Web-based, Participant-driven Studies Yield Novel Genetic Associations for Common Traits." *PLoS Genetics* 6(6): e1000993.

131   "10 hottest nerds": *Newsweek*. 2007. "The 10 Hottest Nerds." *Newsweek* (9 October).

131   "arguably the smartest, most influential": Zimmer, Carl. 2010. "A Day Among the Genomes." *Discover The Loom* Blog (3 May).

132   On the project's website, you can learn: Personal Genome Project. "PGP-10." www.personalgenomes.org/pgp10.html

140   "a race between education and catastrophe": Wells, H.G. 1920 (reprinted 1976). *The Outline of History: Being a Plain History of Life and Mankind*. St. Clair Shores, Mich.: Scholarly Press.

143   "genetic McCarthyism": Green, Robert C., and George J. Annas. 2008. "The Genetic Privacy of Presidential Candidates." *New England Journal of Medicine* 359: 2192–3.

5 Down in the Brain

148   "That genes strongly influence how we act": Holden, Constane. 2008. "Parsing the Genetics of Behavior." *Science* 322(5903): 892–5.

151   "all human behavioral traits are heritable": Turkheimer, Eric. 2000. "Three Laws of Behavior Genetics and What They Mean." *Current Directions in Psychological Sciences* 9: 160–4.

152   "'compulsive hoarding' ... study": Iervolino, Alessandra C., et al. 2009. "Prevalence and Heritability of Compulsive Hoarding: A Twin Study." *American Journal of Psychiatry* 166: 1156–61.

152   your tendency to be religious: Koenig, Laura B., et al. 2005. "Genetic and Environmental Influences on Religiousness: Findings for Retrospective and Current Religiousness Ratings." *Journal of Personality* 73(2): 471–88.

153   holding liberal or conservative values: Alford, John R., et al. 2005. "Are Political Orientations Genetically Transmitted?" *American Political Science Review* 99: 153–67.

153   "new science of human nature": Fowler, James H., and Darren Schreiber. 2008. "Biology, Politics, and the Emerging Science of Human Nature." *Science* 322(5903): 912–4.

153   "most original thinker": McLaughlin, John. 2008. "2008 Year-End Awards." *The McLaughlin Group* (27–28 December). Transcript available at www.mclaughlin.com/transcript.htm?id=697

153   studies of voting behavior: Fowler, James H., et al. 2008. "Genetic Variation in Political Participation." *American Political Science Review* 102: 233–48.

154   a predisposition to homosexuality: Hamer, Dean H., et al. 1993. "A Linkage between DNA Markers on the X Chromosome and Male Sexual Orientation." *Science* 261(5119): 321–7.

155   the "aggression gene": Brunner, Han G., et al. 1993. "Abnormal Behavior Associated with a Point Mutation in the Structural Gene for Monoamine Oxidase A." *Science* 262(5133): 578–80.

156   the gene for the dopamine D4 receptor and … thrill-seeking: Benjamin, Jonathan, et al. 1996. "Population and Familial Association between the D4 Dopamine Receptor Gene and Measures of Novelty Seeking." Nature Genetics 12(1): 81–4. Ebstein, Richard P., et al. 1996. "Dopamine D4 Receptor (D4DR) Exon III Polymorphism Associated with the Human Personality Trait of Novelty Seeking." *Nature Genetics* 12(1): 78–80.

157   In Lesch's study, the subjects with a high neuroticism: Lesch, Klaus-Peter, et al. 1996. "Association of Anxiety-related Traits with a Polymorphism in the Serotonin Transporter Gene Regulatory Region." *Science* 274(5292): 1527–31.

157   Different variants of DRD4 apparently influence … sexual desire: Ben Zion, Itzhak Zahy, et al. 2006. "Polymorphisms in the Dopamine D4 Receptor Gene (DRD4) Contribute to Individual Differences in Human Sexual Behavior: Desire, Arousal and Sexual Function." *Molecular Psychiatry* 11(8): 782–6.

158   gene chips to compare … high IQ and average IQ children: Butcher, Lee M., et al. 2008. "Genomewide Quantitative Trait Locus Association Scan of General Cognitive Ability Using Pooled DNA and 500K Single Nucleotide Polymorphism Microarrays." *Genes, Brain and Behavior* 7(4): 435–46.

161   "Major depression is a familial disorder": Sullivan, Patrick F. 2000. "Genetic Epidemiology of Major Depression: Review and Meta-Analysis." *American Journal of Psychiatry* 157: 1552–62.

165   an analysis of fourteen other studies: Risch, Neil, et al. 2009. "Interaction Between the Serotonin Transporter Gene (*5-HTTLPR*), Stressful Life Events, and Risk of Depression." *Journal of the American Medical Association* 301(23): 2462–71.

168   studied advertisements for SSRI drugs: Lacasse, Jeffrey R., and Jonathan Leo. 2005. "Serotonin and Depression: A Disconnect between the Advertisements and Scientific Literature." *PLoS Medicine* 2(12): 101–6.

171   "Mugged by Our Genes?": Aamodt, Sandra, and Sam Wang. 2009. "Mugged by Our Genes?" *New York Times Opinionator* Blog (24 March). http://

opinionator.blogs.nytimes.com/2009/03/24/guest-column-mugged-by-our-genes

171   Beaver calculated that almost half: Beaver, Kevin M., et al. 2009. "Biosocial Development and Delinquent Involvement." Youth Violence and Juvenile Justice 7: 223–38.

177   imaging studies of our old friend SERT: Hariri, Ahmad R., et al. 2002. "Serotonin Transporter Genetic Variation and the Response of the Human Amygdala." Science 297: 400–3.

186   Eisenberg and Lieberman put their volunteers into a brain scanner: Eisenberger, Naomi, et al. 2007. "Understanding Genetic Risk for Aggression: Clues from the Brain's Response to Social Exclusion." Biological Psychiatry 61: 1100–8.

187   driving ability: McHughen, Stephanie A., et al. 2009. "BDNF Val66Met Polymorphism Influences Motor System Function in the Human Brain." Cerebral Cortex (published online 10 September).

188   slow-release methionine variant ... protects: Pezawas, L., et al. 2008. "MET BDNF Protects Against Morphological S Allele Effects of 5-HHTLPR." Molecular Psychiatry 13(654): 709–16.

188   ninety-five Finnish alcoholics: Sjöberg, Rikard L., et al. 2008. "A Non-additive Interaction of a Functional MAO-A VNTR and Testosterone Predicts Antisocial Behavior." Neuropsychopharmacology 33(2): 425–30.

189   Caspi and Moffitt lab ... on attention deficit/hyperactivity disorder: Caspi, Avshalom, et al. 2008. "A Replicated Molecular Genetic Basis for Subtyping Antisocial Behavior in Children with Attention-Deficit/Hyperactivity Disorder." Archives of General Psychiatry 65(2): 203–10.

191   "The debate about nature versus nurture": Turkheimer, Eric. 2000. "Three Laws of Genetic Behaviors." Current Directions in Psychological Science 9(5): 160–4.

6 *Personality Is a Four-Letter Word*

197   "stable individual differences in the reactivity": Nettle, Daniel. 2007. Personality: What Makes You the Way You Are. Oxford: Oxford University Press, p. 43.

198   Galton went to the dictionaries: Galton, Francis. 1884. "The Measurement of Character." Fortnightly Review 36: 179–85.

198   burrowed through contemporary English dictionaries: Allport, Gordon W., and H.S. Odbert. 1936. "Trait Names: A Psycho-lexical Study." Psychological Monographs 47(211).

199   revealing the five factors: Tupes, Ernest C., and Raymond E. Cristal. 1961. "Recurrent Personality Factors Based on Trait Raitings." Technical Report

ASD-TR-61-97. Lackland Air Force Base, TX: Personnel Laboratory, US Air Forces Systems Command. Norman, Warren T. 1963. "Toward an Adequate Taxonomy of Personality Attributes: Replicated Factor Structure in Peer Nomination Personality Ratings." *Journal of Abnormal and Social Psychology* 66: 574–83.

201   the dynamic duo of Costa and McCrae: Costa, Paul T., Jr., and Robert R. McCrae. 1988. "Personality in Adulthood: A Six-year Longitudinal Study of Self-reports and Spouse ratings on the NEO Personality Inventory." *Journal of Personality and Social Psychology* 54(5): 853–63.

202   individuals … with one of ten categories of clinical: Lisa Saulsman and Andrew Page Saulsman, Lisa M., and Andrew C. Page. 2004. "The Five-Factor Model and Personality Disorder Empirical Literature: A Meta-Analytic Review." *Clinical Psychology Review* 23(8): 1055–85.

202   Conard reviewed the personality scores: Conard, Maureen A. 2006. "Aptitude Is Not Enough: How Personality and Behavior Predict Academic Performance." *Journal of Research in Personality* 40(3): 339–46.

203   conscientiousness and … job performance: Barrick, Murray R., and Michael K. Mount. 1991. "The Big Five Personality Dimensions and Job Performance: A Meta-Analysis." *Personnel Psychology* 44: 1–26.

211   personality is around fifty percent heritable: Bouchard, Thomas J. and Matt McGue. 2003. "Genetic and Environmental Influences on Human Psychological Differences." *Journal of Neurobiology* 54: 4–45.

211   "Behavioural-genetic research provides the best": Plomin, Robert, et al. 2001. "Why Are Children in the Same Family So Different? Nonshared Environment a Decade Later." *Canadian Journal of Psychiatry* 46: 225–33.

211   "Why Are Children in the Same Family So Different": Plomin, Robert, and Denise Daniels. 1987. "Why Are Children from the Same Family So Different from One Another?" *Behavioral and Brain Sciences* 10: 1–60.

212   "The point is that this does not generalize": Nettle, Daniel. 2007. *Personality*, 216.

214   "Understanding genetic mechanisms": Quoted in Holden, Constance. 2008. "Parsing the Genetics of Behavior." *Science* 322(5903): 892–5.

216   "revolutionary mind": *Seed*. 2009. "Revolutionary Minds: The Re-envisionaries." http://revminds.seedmagazine.com/revminds/member/heejung_kim

216   Kim presented a group of Koreans: Kim, Heejung S., et al. 2010. "Culture, Serotonin Receptor Polymorphism and Locus of Attention." *Social, Affective and Cognitive Neuroscience* 5(2–3): 212–8.

217   subjects to act out a job interview: Shalev, Idan, et al. 2009. "BDNF Val66Met Polymorphism Is Associated with HPA Axis Reactivity to

Psychological Stress Characterized by Genotype and Gender Interactions."
Psychoneuroendocrinology 34(3): 382–8.

220   openness is ... linked to cognitive flexibility: Kalbitzer, Jan, et al. 2009.
"The Personality Trait Openness Is Related to Cerebral 5-HTT Levels."
Neuroimage 45(2): 280–5.

220   people ... characterized by *sensory processing sensitivity*: See Aron, Elaine N.
1997. *The Highly Sensitive Person: How to Thrive When the World Overwhelms You.*
New York: Broadway Books.

222   "We conduct this ... to better understand the teenage": IMAGEN Project.
2007. "The IMAGEN Study Has Started at the End of December 2007."
IMAGEN website press release (11 December). www.imagen-
europe.com/en/imagen-study.php

224   Belsky argues we should be thinking ... of *plasticity*: Belsky, Jay, and
Michael   Pluess.   2009.   "Beyond   Diathesis   Stress: Differential
Susceptibility to Environmental Influences." *Psychological Bulletin* 135(6):
885–908.

224   "dandelions" and "orchids": Boyce, W. Thomas, and Bruce J. Ellis. 2005.
"Biological Sensitivity to Context: An Evolutionary-developmental Theory
of the Origins and Functions of Stress Reactivity." *Developmental
Psychopathology* 17(2): 271–301.

225   quality of parental care directly curbs ... the vulnerable: Kinnally, Erin L.,
et al. 2009. "Parental Care Moderates the Influence of MAOA-uVNTR
Genotype and Childhood Stressors on Trait Impulsivity and Aggression in
Adult Women." *Psychiatric Genetics* 19(3): 126–33.

227   "Personal genomics has a long way to go": Pinker, Steven. 2009. "My
Genome, My Self." *New York Times Magazine* (7 January).

228   two American sisters, Tichelle and La'Tanya: Bazelon, Emily. 2006. "A
Question of Resilience." *New York Times Magazine* (April 30).

229   "genetically predisposed neural processing": Schardt, Dina M. 2010.
"Volition Diminishes Genetically Mediated Amygdala Hyperreactivity."
Neuroimage 53: 943–51.

7 *The Interpreter of Biologies*

235   a group led by Mario Fraga: Fraga, Mario F., et al. 2005. "Epigenetic
Differences Arise during the Lifetime of Monozygotic Twins." *Proceedings of
the National Academy of Sciences* 102(30): 10604–9.

237   rat babies ... raised by uncaring mothers: Weaver, Ian C. G., et al. 2004.
"Epigenetic Programming by Maternal Behavior." *Nature Neuroscience* 7:
847–54.

237   tissue from twelve people who had committed suicide: McGowan, Patrick
       O., et al. 2009. "Epigenetic Regulation of the Glucocorticoid Receptor in
       Human Brain Associates with Childhood Abuse." *Nature Neuroscience* 12(3):
       342–8.

238   epigenetic effect … when a woman experiences depression: Oberlander,
       Tim F., et al. 2008. "Prenatal Exposure to Maternal Depression, Neonatal
       Methylation of Human Glucocorticoid Receptor Gene (NR3C1) and Infant
       Cortisol Stress Responses." Epigenetics 3(2): 97–106.

241   the future of the pharmaceutical industry: PricewaterhouseCoopers. 2009.
       "Pharma 2020: Challenging Business Models" (April). www.pwc.com/
       gx/en/pharma-life-sciences/pharma-2020-business-models

248   "We used to think our fate was in the stars": Quoted in Jaroff, Leon J. 1989.
       The Gene Hunt." *Time* (20 March): 62–7.

8 Looking for the New Biological Man

251   attracted to the body odor of men: Wedekind, Claus, et al. 1995. "MHC-
       dependent Preferences in Humans." *Proceedings of the Royal Society London B*
       260: 245–9.

252   Hutterites married following the same rule: Ober, Carole, et al. 1997.
       "HLA and Mate Choice in Humans." *American Journal of Human Genetics*
       61(3): 497–504.

253   "… the hypothesis that these genes influence mate choice": Chaix,
       Raphaëlle, et al. 2008. "Is Mate Choice in Humans MHC-Dependent?" *PLoS
       Genetics* 4(9): 1–5.

254   no HLA preferences … of the Yoruba people: Chaix, Raphaëlle, et al. 2008.
       "Is Mate Choice in Humans MCH-dependent?" *PLoS Genetics* 4(9):
       e1000184. doi:10.1371/journal.pgen.1000184

255   HLA genes of three hundred Japanese couples: Ihara, Yasuo, et al. 2000.
       "HLA and Human Mate cChoice: Tests on Japanese Couples."
       Anthropological Science 108: 199–214.

255   reviews all the experiments done … to mate choice: Havlicek, Jan, and S.
       Craig Roberts. 2009. "MHC-correlated Mate Choice in Humans: A
       Review." *Psychoneuroendocrinology* 34(4): 497–512.

256   difference in their HLA genes … reported a better relationship: Garver-
       Apgar, Christine, et al. 2006. "MHC Alleles, Sexual Responsivity, and
       Unfaithfulness in Romantic Couples." *Psychological Sciences* 17: 830–5.

260   "infidelity gene": Walum, Hasse, et al. 2008. "Genetic Variation in the
       Vasopressin Receptor 1A Gene (*AVPR1A*) Associates with Pair-bonding
       Behavior in Humans." *PNAS - Proceedings of the National Academy of Sciences*
       105(37): 14153–6.

261 gene-tested Mexican couples in San Francisco: Risch, Neil, et al. 2009. "Ancestry-related Assortative Mating in Latino Populations." *Genome Biology* 10(11): R132.

261 "People seem to gauge their partners": Quoted in Aldous, Peter. 2009. "Guapa, It's Your Genetic Ancestry I Love." *New Scientist* (20 November).

264 "Looking down the line ten to twenty years": Parker, Randall. 2010. "Counsyl Genetic Tests for Prospective Parents." *FuturePundit* Blog (2 February). http://www.futurepundit.com/archives/006920.html

265 "The Future of Neo-eugenics": Leroi, Armand M. 2006. "The Future of Neo-eugenics." *EMBO Reports* 7(12): 1184–7.

266 inherited illness has become rarer: Marchione, Marilynn. 2010. "Gene Testing Spurs Decline of Some Dire Diseases." *Associated Press* (19 February).

268 "social entrepreneurs with a mission": Pollack, Andrew. 2010. "Firm Brings Gene Test to Masses." *New York Times* (28 January).

268 "Universal genetic testing can drastically reduce": Quoted in Counsyl. 2010. "Counsyl Test to Prevent Diseases Like Those in 'Extraordinary Measures' Now at 100+ Medical Centers." Counsyl press release (22 January). www.counsyl.com/pr/counsyl-test-to-prevent-diseases-like-those-in-extraordinary-measures-now-at-100- medical-centers

269 compared … African and European genomes: Lohmueller, Kirk E., et al. 2008. "Proportionally More Deleterious Genetic Variation in European than in African Populations." *Nature* 451: 994–7.

271 "sperm donor catalogue … with the most appealing": London Sperm Bank. 2011. "Looking for Donated Sperm?" www.londonspermbank.com/looking_for_donated_sperm.html

271 "The identification of significant risk factors": Goldstein, David B. 2010. "Personalized Medicine." *Nature* 463: 26–32.

272 Danish Abortion Board … refused: Rask Larsen, Julie. 2008. "Dansk abortlov er forældet og krænkende (Danish Abortion Law Is Outdated and Offensive)." *Politiken* (25 August).

273 "Child welfare laws certainly prevent": Appel, Jacob M. 2009. "Mandatory Genetic Testing Isn't Eugenics, It's Smart Science." *Opposing Views* (4 March). www.opposingviews.com/i/mandatory-genetic-testing-isn-t-eugenics-it-s-smart-science

277 "Will 'risk' and 'potential' eventually dominate": Singh, Ilina, and Nikolas Rose. 2009. "Biomarkers in Psychiatry." *Nature* 460 (9): 202–7.

278 the research team from the University of Georgia: Brody, Gene H., et al. 2009. "Prevention Effects Moderate the Association of 5-HTTLPR and Youth Risk Behavior Initiation: Gene x Environment Hypotheses Tested via a Randomized Prevention Design." *Child Development* 80(3): 645–61.

280   "Today's biocriminologies ... are not": Rafter, Nicole. 2008. *The Criminal Brain: Understanding Biological Theories of Crime*. New York: New York University Press, 246.

280   "I want to enlist modern genetics": Ibid., 16.

282   "... a moral response to this question": Lahn, Bruce T., and Lanny Ebenstein. 2009. "Let's Celebrate Human Genetic Diversity." *Nature* 461(18): 726–8.

283   the best way to avoid the problem of politics: Rose, Steven. 2009. "Darwin 200: Should Scientists Study Race and IQ? No: Science and Society Do Not Benefit." *Nature* 457: 786–8.

284   contrasted Asian culture and Western culture: Way, Baldwin M. and Lieberman, Matthew D. 2010. "Is there a genetic contribution to cultural differences? Collectivism, individualism and genetic markers of social sensitivity." *Social Cognitive and Affective Neuroscience* 5(2–3): 203–11.

285   "When enough brains are predisposed": Lieberman, Matthew D. 2009. "What Makes Big Ideas Sticky?" In Brockman, Max (ed.) *What's Next?: Dispatches from the Future of Science*. New York: Vintage, 89–103.

285   "a new science of human nature": Fowler, James H., and Darren Schreiber. 2008. "Biology, Politics, and the Emerging Science of Human Nature." *Science* 322(5903): 912–4.

287   "I'm a conservative reactionary": Hacking, Ian 2009. "Current Controversies: Ian Hacking." *On the Human* Blog (30 March). http://onthehuman.org/2009/03/current-controversies-ian-hacking

Acknowledgments

MY BEAUTIFUL GENOME was for a long time merely an idea and without the generous support that the project has received, it would never have solidified into a book. For this, I wish to express my sincere gratitude to the Danish Arts Council, the Oticon Foundation, and the Carlsberg Memorial Grant.

My thanks go to the many individuals who gave precious time and granted me the chance to explore the personal genomics revolution with them: Linda Avey, 23andMe and the Brainstorm Research Foundation; Jason Bobe, the Personal Genome Project; John Boyce, Consumer Genomics Show; Tamara Brown, GenePartner; Michael Cariaso, BioTeam; George Church, Harvard University and the Personal Genome Project; Earl Collier, deCODE Genetics; Edward Farmer, deCODE Genetics; Anne-Marie Gerdes, Copenhagen University Hospital; Bennett Greenspan, FamilyTreeDNA; Dean Hamer, US National Cancer Institute, National Institutes of Health; Henrik Skovdahl Hansen, Psychological Publishers; Kenneth Kendler, Virginia Commonwealth University, and Sue Kendler; Susanne Kjergaard, Copenhagen University Hospital; Gitte Moos Knudsen, Center for Integrated Molecular Brain Imaging, Copenhagen University Hospital; Jennifer Larsen, H. Lundbeck A/S; Armand Leroi, Imperial College, London; Cecilie Löe Licht, Copenhagen

University Hospital; Diana Gale Matthiesen, Danish Demes; Kirk Maxey; Jen McCabe, Health Management Rx Blog; Ugo Perego, Sorenson Molecular Genealogy Foundation and GeneTree.com; Robert Plomin, King's College, London; Craig Roberts, University of Liverpool; Birgitte Søgaard, H. Lundbeck A/S; Kári Stefánsson, deCODE Genetics; Moshe Szyf, McGill University; Dan Vorhaus, Genomics Law Report; James D. Watson, Cold Spring Harbor Laboratory; Claus Wedekind, University of Lausanne; Daniel Weinberger, US National Institute of Mental Health, National Institutes of Health; Spencer Wells, National Geographic Society; and Scott Woodward, Sorenson Molecular Genealogy Foundation.

For great help and inspiration, I would like especially to thank my contacts at Novo Nordisk, Novozymes, Danisco, H. Lundbeck, and ALK.

I am indebted to Karen Gahrn, Anna Libak, and Thomas G. Jensen for providing valuable comments on the manuscript and to my agent Peter Tallack for assuming the sometimes very difficult task of representing a Danish writer in the English-speaking world. I praise my luck for having had the chance to work with Robin Dennis at Oneworld Publications – now I know what good editing is.

Special thanks go out to Debbie Marks and Chris Sander for their friendship, extraordinary hospitality, and not least for innumerable discussions about science, life, and everything in between.

Finally, my great appreciation to Morten Malling – a truly patient man – for helping to preserve my (relative) mental health.

Index

About the Author

LONE FRANK is the author of *The Neurotourist: Postcards from the Edge of Brain Science*. She holds a PhD in neurobiology and was previously a research scientist in the biotechnology industry. An award-winning science journalist and television documentary presenter, she has written for such publications as *Nature, Science, Scientific American*, and *Frankfurter Allgemeine Zeitung* as well as serving as a staff writer for *Weekendavisen*, Denmark's leading newspaper. She lives in Copenhagen.